Routledge Revivals

Nature's Ideological Landscape

Originally published in 1984 *Nature's Ideological Language* examines the common ideological roots of environmental reclamation and nature preservation. In the general context of European, British and American historical experience, the Jutland heaths of Denmark are taken as a concrete example for a general critique of European and American policy concerning the use of landscape. Two sets of contradictions are highlighted: ideological and practical between development and preservation; and those between scientific, historical aesthetic and recreational motivation for preservation. The book is based on a study of the Jutland heath from 1750 to the present, focusing on the Danish perception of the area as expressed in literary art and in economic journals, topographies and government reports. Against this background, the development of the modern conception of nature is traced and its ideological implications and planning consequences discussed. As a study of humanistic geography, this book will be of interest to geographers, conservationists and planners.

Nature's Ideological Landscape

A Literary and Geographic Perspective on its Development and Preservation on Denmark's Jutland Heath

by Kenneth Olwig

First published in 1984
by George Allen & Unwin, (Publishers) Ltd.

This edition first published in 2019 by Routledge
2 Park Square, Milton Park, Abingdon, Oxon, OX14 4RN
and by Routledge
52 Vanderbilt Avenue, New York, NY 10017

Routledge is an imprint of the Taylor & Francis Group, an informa business

Publisher's Note
The publisher has gone to great lengths to ensure the quality of this reprint but points out that some imperfections in the original copies may be apparent.

Disclaimer
The publisher has made every effort to trace copyright holders and welcomes correspondence from those they have been unable to contact.

A Library of Congress record exists under LCCN: 84002788

ISBN 13: 978-0-367-36971-2 (hbk)
ISBN 13: 978-0-429-35212-6 (ebk)
ISBN 13: 978-0-367-36972-9 (pbk)

NATURE'S IDEOLOGICAL LANDSCAPE

A Literary and Geographic Perspective on its Development and Preservation on Denmark's Jutland Heath

Kenneth Olwig

The Royal Danish Postgraduate School of Educational Studies

London
GEORGE ALLEN & UNWIN
Boston Sydney

**George Allen & Unwin (Publishers) Ltd,
40 Museum Street, London WC1A 1LU, UK**

George Allen & Unwin (Publishers) Ltd,
Park Lane, Hemel Hempstead, Herts HP2 4TE, UK

Allen & Unwin Inc.,
9 Winchester Terrace, Winchester, Mass. 01890, USA

George Allen & Unwin Australia Pty Ltd,
8 Napier Street, North Sydney, NSW 2060, Australia

First published in 1984

ISSN 0261-0485

British Library Cataloguing in Publication Data

Olwig, Kenneth Robert
Nature's ideological landscape.—(London research series in geography; 5)
1. Landscape protection—Denmark—Jutland—History 2. Moors and heaths—Denmark—Jutland—History
I. Title II. Series
719'.09489'5 QH77.D4
ISBN 0-04-71002-8

Library of Congress Cataloging in Publication Data

Olwig, Kenneth Robert.
Nature's ideological landscape.
(The London research series in geography, ISSN 0261-0485; 5)
Bibliography: p. 104
Includes index.
1. Landscape protection—Denmark—Jutland—History.
2. Reclamation of land—Denmark—Jutland—History.
3. Land use—Denmark—Jutland—Planning—History.
4. Moors and heaths—Denmark—Jutland. 5. Jutland (Denmark)—History. I. Title. II. Series.
QH77.D4048 1984 333.73'16'094895 84-2788
ISBN 0-04-710002-8 (pbk.)

Set in 10/12 point Bembo by Inforum Ltd, Portsmouth

Preface and acknowledgments

This study traces the common ideological roots of two apparently contradictory activities: environmental reclamation and nature preservation. Taking its point of departure in the general context of European, British and American historical experience, it focuses on the case of the reclamation and preservation of the Jutland heaths of Denmark. This concrete example provides a basis for a general critique of European and American policy concerning the use of landscape. It is especially directed to the ideological and practical contradictions between development and preservation, particularly in underdeveloped regions, and the contradictions between scientific, historic, aesthetic and recreational motivations for preservation.

The basis for this book is an historical geographical study of the Jutland heaths from *c.*1750 to the present, which focuses on the Danish perception of the area as expressed both in literary art and in non-artistic sources such as economic journals, topographies and government reports. The development of the modern conception of nature, its ideological implications and its consequences for planning is traced against this background. The book might be termed "applied humanistic geography", because it combines an analysis of art as a vehicle bearing and shaping ideas and attitudes concerning nature and landscape with a study of the practical policy making which transforms landscape and provides the parameters for its development, thus providing an empirical reference for a critique of modern landscape development and planning practice.

I would like to thank those who have commented at length on the manuscript at different stages in its development: Denis Cosgrove for his loyal, yet critical editorship; Clarence Glacken for his eye for the strengths in a manuscript; David Lowenthal and Karen Fog Olwig for their constructive criticism; Yi-Fu Tuan for being an academic catalyst who is always open to the ideas of others and never imposes his own.

I would then like to thank those who have read and discussed this or earlier versions of the manuscript with me. They are: Søren Baggesen, Viggo Hansen, Orvar Löfgren, A. Karup Mogensen, Robert A. Olwig and Hans Vejleskov.

This manuscript has undergone many changes *en route* to the publisher and so I must take full responsibility for its deficiencies.

Finally, I would like to thank the Danish State Research Council and the Danish Marshall Fund for helping to make my research possible.

The following have kindly granted permission to quote from published works:

J. M. Dent & Sons, from Virgil, *Eclogues and Georgics*, T. F. Royds (trans.).
American-Scandinavian Foundation, from Steen Steensen Blicher, *Twelve stories*, Hanna Astrup Larsen (trans.).

Illustrations are acknowledged individually.

Contents

List of figures

Introduction

It is common today to think of nature as wilderness landscape such as we find in the Yellowstone or Yosemite national parks in the United States. However, the term nature is also used in another sense which does not appear to be related to landscape, namely to refer to what is natural as opposed to what is unnatural. In this sense nature refers to human values rather than to a material entity. The two senses of the word are related, however. Historically, this relationship is best understood by examining the transformation which the word nature has undergone since classical times, from a normative conception of what the process of social and environmental development ought to be to a positive conception of a landscape type.[1] Initially it is difficult to see what the two ends of this historical transformation have to do with one another. How is nature as wilderness landscape – something which we protect from development – related to nature conceived as a cosmological principle of development?

One thing which the classical and modern conceptions of nature have in common is their ideological importance. The classical conception entailed normative values concerning the process of development through stages of increased reciprocity between people in society and between society and its environment. Its ideological import lay in its expressing the characteristic assumptions of classical thinkers about the origins and future of their world. When we use nature to refer to wilderness landscape the connection between nature and ideology is less obvious, but in fact nature as landscape is strongly identified with norms and values in our society. The preservationist Edward Abbey, for example, places wilderness parks in the same category as other monumental symbols of society's values:

> We have agreed not to drive our automobiles into cathedrals, concert halls, art museums, legislative assemblies, private bedrooms, or other sanctums of our culture. We should treat our National Parks with the same deference . . .[2]

In fact, nature often *is* treated with the same deference as other sanctums of culture. Human artifacts and natural features are thus protected under the same legislation as "national monuments" in the United States and as "nature" in Denmark. But whereas existing religious practices and political institutions are directly linked with the values identified, for example, with cathedrals and legislative assemblies, the precise values associated with nature's landscape are less apparent. This does not necessarily mean that such landscape has no social impact. Indeed the life and economy of the peripheral regions, where such wilderness is commonly found, can be substantially affected by the actions aimed at preserving them.

This book examines the relationship between nature's landscape and ideology, and explores the impact of an ideology of nature on regional development. It seeks to do this through historical examination of changes in Western concepts of nature and their relationship with the use of environment. The study of literature and art plays an important role in this analysis because the concept of nature is intricately bound up with other cultural values, particularly our notion of the beautiful. The focus of the study is the historical geography of a particular region. Grand abstractions like Western society or the concept of nature are focused and sharpened when they are seen in the light of actual societies with real landscapes and a real body of art and literature.

The Jutland heath

The heathlands of Danish Jutland are used here as a case study to elucidate the relationship between changes in our concept and our use of nature. A number of reasons make this a suitably representative area upon which to concentrate. The Jutland heaths are noteworthy for the intensity of development which they have experienced and the relationship between that development and national ideology. They are, perhaps paradoxically, noteworthy too for the intensity of the movement of their preservation. Jutland can be regarded as a model both of regional development and nature preservation. More generally, heathlands are significant as landscape because they have provided a landscape type of great attraction for European interest in nature.[3] They characterize many peripheral regions of Europe, for example highland Scotland and parts of Ireland. Conclusions drawn from the Jutland study will therefore be pertinent to other such areas. Moreover, Denmark provides a manageable microcosm of European society, drawing influence from both the continent and the British Isles. With a small population and limited political influence Denmark has been culturally and economically involved with the outside world to a degree which larger, more self-sufficient, societies have not. Consequently, the story of the Danish heaths must continuously refer to ideas and developments occurring elsewhere in European and, by extension, American culture.

The study of literary texts provides a tool for elucidating the ideas and values attached to wilderness landscape and nature. This extends beyond the literature directly identified with Jutland to the wider European heritage in which the local literary tradition is embedded. Further, literature is not merely illustrative; it provides a point of departure in that much of the changing meaning of nature is explored and articulated in imaginative literature – many of the environmental alterations in Jutland were promoted in poetry and the novel. Therefore considerable use is made of excerpts from texts, both fiction and non-fiction.

The heath

Webster's Dictionary defines a heath as:

> A tract of wasteland . . . an extensive area of rather level open uncultivated land usually with poor coarse soil, inferior drainage, and a surface rich in peat or peaty humus.

Its characteristic plant is heather, which is described as:

> Any of a family (Ericaceæ the heath family) of shrubby dicotyledonous and often evergreen plants that thrive on barren usually acid and ill-drained soil; especially an evergreen subshrub of either of two genera (*Erica* and *Calluna*) with whorls of needlelike leaves and clusters of small flowers.[4]

These definitions tell all, yet tell nothing, about the heath and its characteristics. Words like "poor coarse soil," "inferior drainage," and "small flowers," are hopeless as aids to our understanding the attraction of the

Figure 0.1 Heathscape from *Bærentzens Danmark*, 1855.

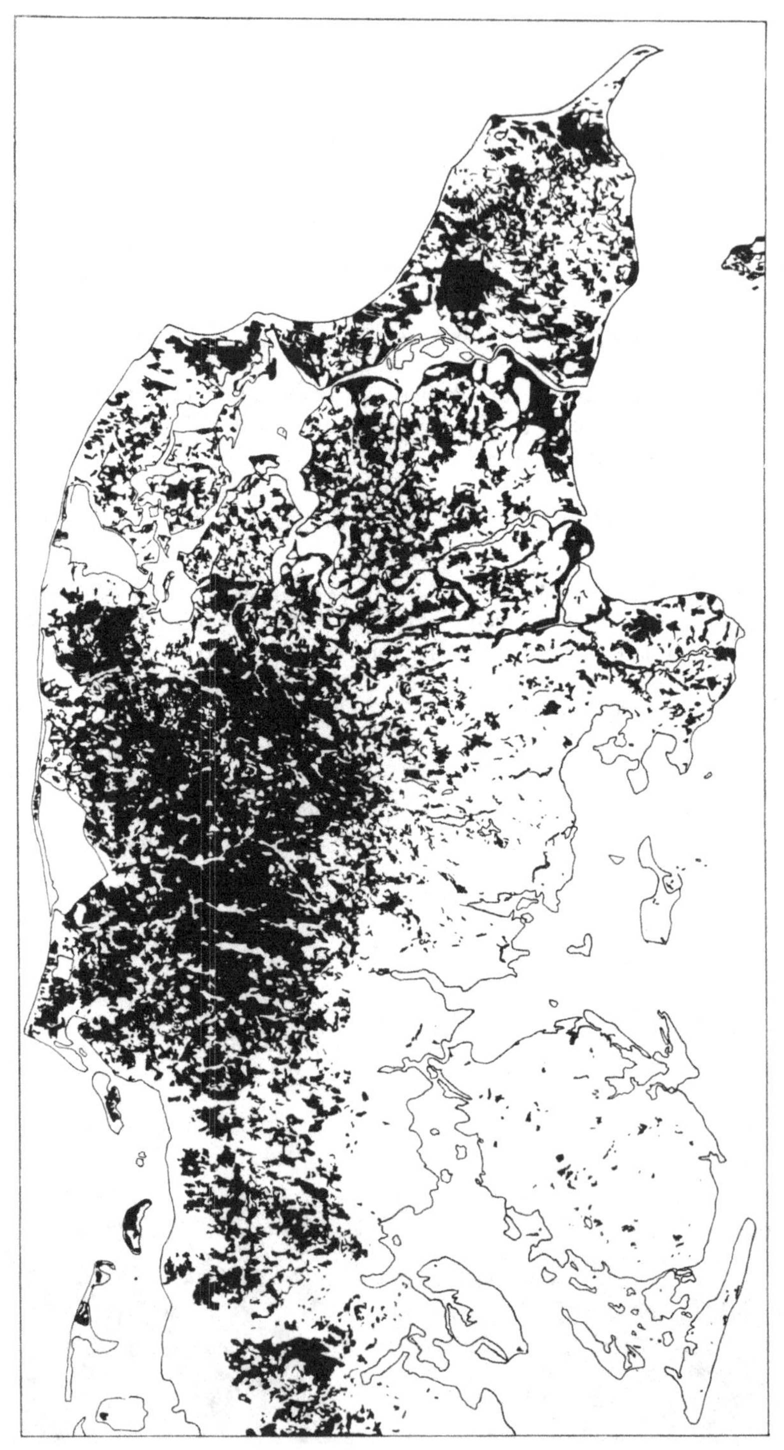

Figure 0.2 The extent of the Jutland heath about 1800 (map prepared by the University of Copenhagen's Geographical Institute and based on the Royal Society of Sciences & Letters' map of 1762).

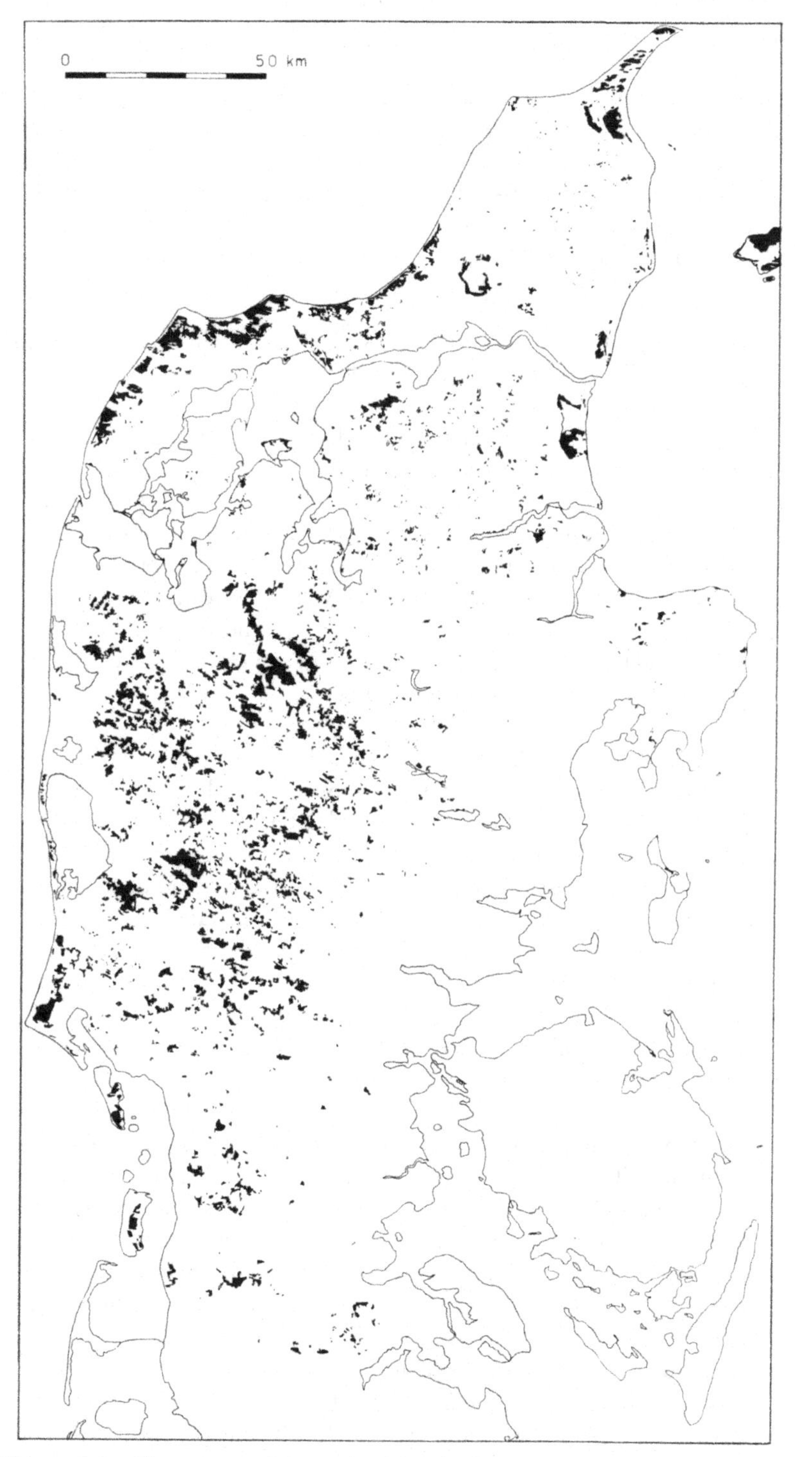

Figure 0.3 The extent of the Jutland heath about 1950 (map prepared by the University of Copenhagen's Geographical Institute).

heath to Shakespeare, Burns, John Clare or Thomas Hardy. Other European writers, not least Danes, have expressed a love of the heathlands.[3] Using the Danish example as my point of departure and drawing on European, particularly British, literature and history, I shall try to explain just what it is about those small flowers, that aromatic sea of late summer purple and those needlelike leaves, turning almost black in winter, that gives heath nature its compelling attraction.

Similarly, standard geographical descriptions of the Jutland heaths' development tell everything, yet nothing:

> By far the greatest concentration of heath was to be found in Jutland, particularly the west and north, where it formed a characteristic landscape covering approximately three million acres [1.2 million ha] in 1800, or about 40 per cent of the entire area of the peninsula . . . With the foundation of the Danish Heath Society in 1866, with its headquarters first in Aarhus and later at Viborg, a really determined attack on the heath began. By 1950 the extent of heath, dune and bog in Jutland had been reduced to 640,000 acres [259 000 ha] representing only 8.8 per cent of the peninsular area.[5]

> The demand for new land rose, especially after the loss of north Slesvig in 1864. Bravely and industriously, the local population waged war on the heather, encouraged after 1866 by the Danish Heath Society (Det danske Hedeselskab). How successful the reclamation has been is testified by the well laid out farms and plantations which now cover the outwash plains and the adjacent poorer moraines. After a century, over four-fifths of the heathland have been converted to agriculture and forestry . . . Only a few patches of heath remain as scar-like reminders of the former waste, but in several places small areas of heath have been carefully preserved lest future generations forget the labours of the past.[6]

Such descriptions focus on the physical character of the heath and its transformation. They neglect those aspects of the heath that first made its cultivation a patriotic national cause, and that later justified its preservation as equally patriotic. Equally, standard explanations of changing use that appeal to economic necessity and political expediency are, as we shall see, inadequate. Rather, the development and later preservation of the Danish heath must be viewed in the light of a complex ideology which motivated both nationalist movements and artistic activity. By extension, Denmark's location within a broader European history allows what follows to shed light on the role played by heath landscapes in the changing interpretation of nature in the West.

Notes

1 It is particularly dangerous to generalize about the word nature. Raymond Williams terms it "perhaps the most complex word in the language," and Arthur O. Lovejoy lists 66 meanings under the heading "some meanings of 'nature' ". I have focused upon the transition in the word's meaning from normative process to landscape because it is most relevant to my geographical theme, and in the interest of clarity I have preferred concrete example to semantic argument. Useful concise etymological studies of the word can be found in standard dictionaries (OED, Merriam-Webster), and in: Lovejoy and Boas (1935), *Primitivism and related ideas in antiquity*, pp. 447–56, Williams (1972), Ideas of nature, *Ecology, the shaping enquiry* pp. 146–64, Williams (1976), *Keywords*, pp. 184–9.
2 Abbey (1971), *Desert solitaire: a season in the wilderness*, p. 60: quoted in Graber (1976), *Wilderness as sacred space*, p. 77.
3 Hard (1965), Arkadien in Deutschland; Bemerkungen zu einem landschaftlichem Reiz, pp. 21–41; and Shoard (1982), *The lure of the moors*, pp. 59–73.
4 *Webster's seventh new collegiate dictionary* (1963): "heath," "heather."
5 Thorpe (1957), A special case of heath reclamation in the Alheden district of Jutland, 1700–1955, p. 87.
6 Fullerton and Williams (1972), *Scandinavia*, pp. 111–13.

English language readers are referred to the above two works for a detailed geographical description of the heathlands. Note that the author uses the Danish spelling for "Slesvig", rather than the German "Schleswig"; and the Danish "Holsten", rather than the German "Holstein". A useful history of Denmark can be found in Oakley (1972), *A short history of Denmark*. Danish language readers can find a concise geography of the heaths in Hansen (1970). Hedens Opståen og Omfang, pp. 9–106.

All translations from Danish are by the author unless otherwise noted.

1 *The morphology of the ideological landscape*

From classical times to the Enlightenment, nature had meaning primarily as process: a norm, or principle, of development. This process lay at the core of social values, values relating to social interaction and to interaction between society and its environment. The concept of nature as a thing is, by contrast, characteristically a product of the modern age. But the contemporary idea of nature can be regarded as having developed out of the classical conception and to have carried with it, in transmuted form, many of those former values. Paradoxical as it may seem, the modern urge to protect nature from development, to create stasis, grows out of a conception of nature as change. An examination of the literary appropriation of nature will indicate how this occurred.

The classical conception of nature

Writing "Of classical landscape," in *Modern painters*, John Ruskin was struck by the difference between the classical and modern conceptions of natural beauty:

> [classical artists] shrank with dread or hatred from all the ruggedness of lower nature – from the wrinkled forest bark, the jagged hill-crest, the irregular, inorganic storm of sky; looking to these for the most part as adverse powers, and taking pleasure only in such portions of the lower world as were at once conducive to the rest and health of the human frame, and in harmony with the laws of its gentler beauty.
>
> Thus, as far as I recollect, without a single exception, every Homeric landscape, intended to be beautiful, is composed of a fountain, a meadow, and a shady grove.[1]

By Ruskin's day it had become conventional to think of nature as wild and rugged, primeval and original. Nature was something other than, and opposed to, society, and it was from this otherness that the taste for the wild derived its impetus. The classical writer, however, had preferred a tamed, humanized or "pastoral" landscape of fountain, meadow and shady grove. Human society is very much implicated in the classical landscape.

What seems to be involved here is a change in taste for natural landscape. In fact, one can hardly speak of a mere change in state because classical writers neither conceived of nature as an entity, a landscape, nor even possessed a concept of landscape.[2] We are dealing therefore with neither a development in taste, nor a simple matter of semantics, but with a fundamental shift in social and environmental values, a shift which ultimately is related to alterations in social relations and environmental use.

R. G. Collingwood argues that the word "nature" in classical times referred primarily to a cosmological principle of "development, growth or change". This developmental process "takes successive forms α, β, γ, . . ., in which each is the potentiality of its successor". It thus exhibits, "an eternal repertory," in which the items are "related logically, not temporally, among themselves".[3] This conception is also indicated by John Passmore's etymological analysis of nature:

> The word "nature" derives, it should be remembered, from the latin *nascere*, with such meanings as "to be born," "to come into being". Its etymology suggests, that is, the embryonic, the potential rather than the actual. We speak in this spirit, of an area still in something like its original condition as "not yet developed". To "develop" land, on this way of looking at man's relationship to nature, is to actualize its potentialities, to perfect it.[4]

The modern conception of nature as wilderness is related to the classical in so far as wilderness is seen as being "primeval," or "original," – by implication not yet developed. There is an important difference, however. Whereas in Passmore's terms, wilderness represents an original state, the classical mind applied the term nature not to an actuality, but to a principle of "actualization," an embryonic potential always ready to be realized. Thus, each stage of natural development was seen as an original state, because each was present in embryo within its predecessor. Whereas the modern conception is of nature as a thing, such as a primeval, untouched wilderness, in classical thought nature was the potential governing an eternal repertory of the life cycle of organisms and seasons. Birth, or spring, is embryonic of maturity, or summer, which in turn is embryonic of old age, or autumn, and in turn, the death of winter holds the potential of generational, or seasonal, rebirth. In the classical conception, wilderness was not a thing to be preserved from development, but rather an undeveloped potential.

We still use the word nature in something close to the classical sense in statements like: "it is the *nature* of a being (or it is *natural* for a being) to be born, mature, have children, grow old, and die." This sense readily extends to the realm of values, as in the statement, "it is not the nature of children to have children and hence it is unnatural for children to engage in sexual intercourse". Here, the normal, or natural, course of the generative cycle forms the basis for sexual values. It is the identification with such fundamental values which makes nature, in A.O. Lovejoy's words, "the chief and most pregnant word in the terminology of normative provinces of thought in the West".[5] However, the continued existence of normative values associated with nature tends to be obscured to the degree that nature has become a thing, a landscape, rather than a principle. The argument for preserving wilderness can therefore become tautologous: because wilderness is natural, it is natural to preserve it. The values attached to nature preservation become clearer when the modern concept of nature as landscape is examined in the light of the classical conception that has inspired so much of later European and American thought.

Virgil and nature's landscape

The classical conception is nowhere more apparent than in the writings of Virgil, the Mantuan poet (70–19 BC) whose verse subsequent generations have regarded as a classic literary model and to which Europeans have returned again and again. Virgil provides a point of reference to which we can turn in order to show how succeeding writers have reinterpreted the ideas embodied in his work.[6]

When Ruskin claimed that for the classical mind every "landscape, intended to be beautiful, is composed of a fountain, a meadow, and a shady grove" he was referring to a landscape commonly believed to represent the first stage in social development: the pastoral.[7] The pastoral society of Arcadia, home of the earliest men, was characterized, according to Polybius (?205–?125 BC) by "common assemblies and sacrifice . . . in common for men and women, and choruses of girls and boys together". These collective activities educated people in social life and had the purpose of "smoothing and softening" the effects of a hard environment.[8] In such society, Virgil claimed there was "no fence or boundary-stone to mark the fields," because there was no private ownership and men laboured to replenish a common store.[9] The socialization of individuals into a community parallels the socialization of the environment. This too is smoothed and softened into a communally owned, open, grassy environment of meadow, fountain and grove. Although grazing makes extractive use of the environment, unlike hunting it requires active domestication of animals and land, and thereby humans enter into a reciprocal relationship with the material environment. The meadow is created or extended through flooding, the removal of forest cover and brush, periodic burning, the planting of grass, mowing and, of course grazing. Damming a spring creates a pool or fountain to water the meadow and to provide drink for thirsty livestock and shepherds. The grove is what remains of the primeval forest after it has been thinned, the lower branches of trees trimmed by the pastoralist and his flock. This environment, nestled in the hills and cooled by gentle zephyrs, presents an elemental harmony of earth, wind, fire and water. It stands in stark contrast to the elemental extremes of wilderness with its "wrinkled forest bark, jagged hill-crest, and irregular, inorganic storm of sky," not to mention the dry expanse of desert, or its marine sibling, the ocean. In such environments one element or another dominates and the balance and harmony of the pastoral world is absent.[10] Because the pastoral stage is the natal state of society, like the childhood of the individual it is characterized by a high degree of dependency and passivity. The pastoralist modifies the environment but does not reciprocate with the intensity of the agriculturalist. He does not plow and fertilize. The shepherd's classical posture is one of recline under a shade tree, playing the pipes and singing to a nubile companion, while his livestock graze peacefully. Love suffuses a sheltered, fertile and womblike environment whose deity is the autochthonous Pan; the same passive dependency characterizes pastoral poetry. The poet presents himself as a shepherd; his lyric verse is a direct expression of his participation in pastoral activity. Just as the pastoralist's livelihood and religion are drawn directly from the earth, so his environment becomes the poetic realm, the vehicle of poetic expression. Environment in this poetry does not manifest itself as landscape or scenic backdrop, but rather as an integral, unreflected, part of the poet's being.[11]

To the classical mind, the pastoral environment was an expression of the nature of pastoral society. It is in this *natal* state that the *nature* of the *nation* is formed; the three words share the same Latin root.[12] The suckling babes, the shepherd twins Romulus and Remus, symbolize the Roman nation whose founding in a landscape of "fountain, meadow and shady grove" justified the special place of that landscape in classical lore. As a repository of the innocent virtues of society's childhood, the pastoral stage and its landscape incorporated what were regarded as natural social values. The virtues of the shepherd founders of Rome were necessarily held up to each generation as a model. By the same token the characteristic elements of the pastoral environment, such as the shade tree, were later endowed with the virtues of cohesion, generational fertility and continuity identified with society's pastoral stage:

> But the guardian tree
> Is deeplier buried in the dark earth's womb;
> . . .
> No wind or swollen rain, shall rend it thence:
> Stablished it is, and as the years roll on
> Sees many a son and grandson pass away,
> . . .
> Its brawny arms are spread, and from the midst
> Itself sustains a firmament of shade.[13]

Critics often contrasted the natural, ideal qualities of pastoral society with what they regarded as unnatural deviations from normal social development in Roman society. Such critical comparison was an important function of pastoral poetry, as George Puttenham observed in 1589:

> the Poet devised the *Eglogue* . . . not of purpose to counterfait or represent the rusticall manner of loves and communication: but under the vaile of homely persons and in rude speeches to insinuate and glaunce at greater matters.[14]

Not only shepherds, but the pastoral environment itself was used as a vehicle for social comment. Virgil employed it in the first *Eclogue*, to make a biting comment upon the unnatural policies of Augustan Rome, which in the poem is symbolized by settlement of strangers on the shepherd's native soil. The unnaturalness of this state is suggested by his images of the pastoral shade tree, blasted by lightning, the pastures covered with stone, springs and fountains becoming dry. The coming of strangers and the response of the environment through infertility forced the shepherd and his emaciated flock into exile.[15] The impact of such poetry upon public policy is uncertain, but Bruno Snell has argued that at an ideological level it was not inconsiderable:

> After the disastrous anarchy of the civil wars the desire for peace was paramount, especially among the better minds of the age. Thus Virgil's poetry reflects a genuine political reality, and it is not without significance that Virgil, at a time when Augustus was only just beginning to make his authority felt in the affairs of Rome, had already voiced that yearning for peace which August was fated to satisfy. In this sense Virgil may have been said to have determined to a considerable extent the political ideology of the Augustan age, and his *Eclogues* did indeed exercise an important political and historical function.[16]

While the pastoral stage of society possessed the ideal qualities of the nation's birth and could be held up as a model of natural social behavior, it was equally natural for this embryonic stage to develop into another. The second stage was one of cultivation in which humanity is far less passive and dependent, but engages in a more active reciprocity with the environment. Virgil indicates the beginning of the next stage by concluding the *Eclogues* with the admonishment, "let us rise, it is late, and shade is harmful to crops".[17] But his recognition that this is a natural sequence of development does not prevent Virgil from mourning the loss of the pastoral stage. To do so is as natural as the wistful glance back to childhood innocence. Virgil's verse in the *Eclogues*, regretting the death of pastoral innocence, itself became a primary source for the elegiac genre in poetry. Death also inscribes itself into the pastoral paradise on a skull or tomb: "*Et in Arcadia Ego*," and symbols of death became a requisite in late Renaissance paintings of pastoral scenes.[18]

The nature of the agricultural stage of society

In the *Georgics*, Virgil describes how Jove stimulated humanity to rise up from pastoral passivity and take control of its existence:

> No unlaborious path the Father willed:
> He first taught method as the means, and spurred
> The wits of men by cares, and suffered not
> His realms to slumber 'neath inveterate sloth.
> Before Jove's reign no farmers tilled the soil;
> No fence or boundary-stone to mark the fields
> Religion sanctioned: to the common store
> All labour tended, and the earth herself
> Gave all more freely for that no man asked.
> Then Jove endowed that cursed thing, the snake,
> With venom, and the wolf with thirst for blood,
> Lashed the still sea, shook honey off the trees,
> Robbed men of fire and emptied river-beds
> Which flowed apace with wine; to make men prove
> And hammer out by practice divers arts;
> Now slowly learning how to plough and sow,
> Now striking from flint-vein the lurking fire.
> Then sailors told the number of the stars.
> Toil conquered all, Unconquerable Toil, and Poverty,
> The spur of hungry men . . .
> Taught men the use and method of the plough.
> Soon corn received its special plagues: the stalks
> Were gnawed by mildew, and the thistle reared
> Its head of sloth: death takes the crops, up comes
> A mass of matted undergrowth
> . . .
> So ply your hoes and give the weeds no peace
> . . .
> Now what weapons hardy rustics need
> Ere they can plough or sow the crop to come.[19]

This passage expresses environmental values at the heart of the classical conception of nature. Passmore claims that "to 'develop' land, on this way of looking at man's relationship to nature, is to actualize its potentialities, to perfect it". The focus of Virgil's interest, however, is not the development of land so much as the development of society which, by acquiring the art of cultivation, is able to actualize its own potential. The natural development is reflected through the transformations which the environment undergoes through interaction with society. The environment *may* revert to an earlier state, but to do so is unnatural and contrary to normal social evolution. In the following passage, Varro (166–27 BC) remonstrates against contemporary Roman imperialism for its effects on agricultural practice:

> It was not without reason that those great men, our ancestors, put the Romans who lived in the country ahead of those who lived in the city . . . As therefore in these days practically all the heads of families have sneaked within the walls, abandoning the sickle and the plough, and would rather busy their hands in the theater and in the circus than in the grain-fields and the vineyards, we hire a man to bring us from Africa and Sardinia the grain with which to fill our stomachs, and the vintage we store comes in ships from the islands of Cos and Chios.
>
> And so, in a land where the shepherds who founded the city taught their offspring the cultivation of the earth, there, on the contrary, their descendants, from greed and in the face of the laws, have made pasture out of grain lands – not knowing that agriculture and grazing are not the same thing . . . For grazing cattle do not produce what grows on the land, but tear it off with their teeth.[20]

Reversion from the agricultural state to the pastoral is unnatural because it involves a return to a less reciprocal relationship with the environment. The husbandman cares for the land, the cattle merely take its goodness. Unlike the collectively organized shepherds who established the Roman nation, the individual greed of the present generation flies in the face of collectivity.[21] Such asocial behavior reverts back beyond the pastoral to a savage, inhuman state of self-interest. The characteristic elements of this pre-social stage are described succinctly by Lucretius (?96–55 BC) in *The nature of things*:

> Through many decades of the sun's cyclical course they [mankind] lived out their lives in the fashion of wild beasts roaming at large. No one spent his strength in building the curved plough. No one knew how to cleave the earth with iron, or to plant young saplings in the soil or lop the old branches from tall trees with pruning hooks. Their hearts were well content to accept as a free gift what the sun and showers had given and the earth had produced unsolicited . . .
>
> They could have no thought of the common good, no notion of the mutual restraint of morals and laws. The individual, taught only to live and fend for himself, carried off on his own account such prey as fortune brought him.[22]

Given the cyclical conception of nature as "eternal repertory," in which each stage contains the seeds of its successor, men were justified in fearing that society would ultimately dissolve into such a savage state. Cyclical return to a state of chaos would be natural in the long run, but unnatural if it occurred prematurely. Virgil clearly felt that such premature reversal was well under way:

> . . . right and wrong unseat each other,
> . . .
> Wars roll unceasingly and wickedness
> Assumes a thousand faces; to the plough
> Due honour is denied; fields lie unkept,
> For war has stolen the husbandmen away,
> And straightened all their sickles into swords.[23]

The fear of an incipient return to savagery underlay the "dread and hatred" of wild environments which Ruskin regarded as characteristic of the classical mind.

Within nature's eternal repertory the age of innocence and perfection would return once again after the period of chaos and dissolution. Virgil predicts exactly this in his fourth *Eclogue*, in which the rebirth of society is symbolized by the birth of a child who will lead this second golden age:

> Now the last age of Cumaen song has come. The great line of the ages is born anew. Now the Virgin [justice] returns, the kingdom of Saturn returns. Now a new race is sent down from heaven above. Only do you, chaste Lucina, look benignly on the newborn boy, at whose coming the iron race shall first cease and a golden race will spring up in the whole world. . . . But for you, child, the earth will pour forth with no tilling, as its first gifts, ivy wandering everywhere about with foxglove, and the bean plant mingled with smiling acanthus. Of their own accord goats will bring home udders swollen with milk, and the herds will not fear mighty lions. Of itself your birthplace will pour harmless flowers. The snake, too, will die, and the deceitful poison herb will die. Assyrian nard will grow everywhere. But as soon as you can read the glories of heroes and the deeds of your father and comprehend the meaning of courage, little by little the field will grow yellow with soft corn and the reddening grape will hang from uncultivated bramble bushes, and hard oaks will drip dewy honey.[24]

This passage is important not just because it illustrates regeneration within the natural process of development, but also because of its similarity to the Christian conception of a return to an earlier paradisical state through the birth of a child. This accounts in part for Virgil's acceptance by the medieval Church as a minor Christian prophet. It was largely through

Virgil that Christian and classical notions of past perfection and future possibility could be reconciled. Virgil's work allowed for a convergence of separate mythological traditions within literature.[25]

Nature in time and space

Classical nature was not primarily conceived as a temporal process; its items were related logically. This means that the stages of development could occur simultaneously in different parts of space. In the *Georgics*, both pastoral and agricultural society co-exist with, and at relative distances from, urban society. It is possible to follow the natural process of development spatially from pastoral periphery through cultivated zones and finally to the city. The same space/time logic occurs in the *Eclogues*, which not only refer to a pastoral golden age, distant in time, but also locate the pastoral Arcadia on the upland fringes of classical society. In this countryside the goddess of justice is said to have left her last footprints, and here vestiges of the ideal qualities of earlier times are preserved.[26] Natural qualities could thus be found by moving out in space as well as back in time.[27]

Nature's ideological landscape

The classical conception of nature may be regarded as ideology in the broadest sense, as the content of thinking characteristic of an individual, group, or culture.[28] Classical nature is central to our understanding of classical thought because through it people made sense of their world and its objects. Virgil's poetry, as we have seen, embodied and gave powerful symbolic form to that concept of nature. It is this which accounts for its impact in the Augustan age. In the following chapters we will explore the ways that this concept of nature and its ideological role have been transformed, and how that transformation bears upon society's changing perception and use of the Jutland heath as an ideological landscape.[29]

Notes

1 Ruskin (1904), Of classical landscape: in *The Works of John Ruskin*, vol. IV, p. 234.
2 Parry (1957), Landscape in Greek poetry, pp. 3–29.
3 Collingwood (1976), *The idea of nature*, pp. 43–8.
4 Passmore (1974), *Man's responsibility for nature*, p. 32.
5 Lovejoy (1927), "Nature" as aesthetic norm, p. 444.
6 Eliot (1944), *What is a classic?*
7 Glacken (1956), Changing ideas of the habitable world: in *Man's role in changing the face of*

the Earth, pp. 70–92. Also Glacken (1967), *Traces on the Rhodian shore*, pp. 130–47.

8 Polybius, *Historia* IV: 19–21: quoted in Lovejoy and Boas (1935), *Primitivism and related ideas in antiquity*, pp. 345–8.

9 Virgil *Georgics* I: 151–2: translated in Virgil (1946) *Eclogues and Georgics* (T. F. Royds, trans.), p. 69.

10 On the classical conception of man and the elements, see Glacken (1967), pp. 80–115.

11 Parry (1957), pp. 3–29.

12 Lewis (1967), *Studies in words*, p. 25.

13 Virgil *Georgics* II: 347–57: in Virgil (1946), p. 105.

14 Puttenham (1589), *Arte of English poesie*: quoted in Barrell and Bull (1974), *The Penguin book of English pastoral verse*, p. 21.

15 Putnam (1970), *Virgil's pastoral art*, pp. 20–81. On the negative image of the pastoral, see Marx (1964), *The machine in the garden*, pp. 16–24; and Williams (1973), *The country and the city*, pp. 13–34.

16 Snell (1953), *The discovery of the mind*, p. 291.

17 Virgil, *Eclogues* X: 74–7: quoted, translated and interpreted in Putnam (1970), pp. 386–94.

18 Panofsky (1955), *Meaning in the visual arts*, pp. 295–320.

19 Virgil, *Georgics* I: 147–94: in Virgil (1946), pp. 69–71.

20 Varro (1960), *On Agriculture* (W.D. Hooper, trans.; H.B. Ash, revised), p. 291.

21 Glacken (1967), pp. 116–49: and Tuan (1976), Geopiety: a theme in man's attachment to nature and to place, 11–39.

22 Lucretius (1951), *On the nature of the Universe* (R. E. Latham, trans.), 199–200. Lucretius' characterization of the wild man reflects the classical disdain for asocial characteristics. From Lucretius's time to Rousseau's it was common to perceive ideal qualities amongst primitives, but these primitives normally belonged more to the pastoral state than to the wild. Their environs need not be the comfortable environment of fountain, meadow and shady grove described by Ruskin. On this subject, see Lovejoy (1923), The supposed primitivism of Rousseau's *Discourse on Inequality*, pp. 165–86; and Lovejoy and Boas (1935), *Primitivism and related ideas in antiquity*.

23 Virgil, *Georgics* I: 594–9: in Virgil (1946), p. 88.

24 Virgil, *Eclogues* IV: 4–34: quoted, translated and interpreted in Putnam (1970), pp. 136–65. According to Collingwood, the cyclical notion of the natural process was the most common, as certainly is the case in this influential passage from Virgil. Other variations on this theme, however, were possible. See Lovejoy and Boas (1935), pp. 1–7. On the role of cycles in literature, see Frye (1971), *Anatomy of criticism*.

25 Curtius (1953), *European literature and the Latin Middle Ages* (W. R. Trask, trans.), pp. 18–19, 36.

26 Virgil, *Georgics* II: 564–65: in Virgil (1946), p. 114.

27 On the spatial aspects of the pastoral, see Marx (1964), pp. 3–33. On the classical idealization of places distant in space, see Lovejoy and Boas (1935).

28 *Webster's seventh new collegiate dictionary* (1963), "ideology."

29 The concept of landscape emerged in the Renaissance to describe paintings of inland "natural" scenery – a new artistic vogue. These scenes were not so much of actual places as imagined scenes of places described in the Bible and classical literature – not least that of Virgil. It was also at this time that the word nature was first applied to scenery. In this way the ideology identified with the classical concept of nature was transferred to landscape. Landscape soon ceased to be an exclusively esthetic category. It could be studied from the perspective of physical science, as physical geographers do, or as an expression of culture, as is the case with human geographers (see Ch. 3).

2 *The nature of Gothic Jutland*

Virgil's *Eclogues* begin by describing the destruction of the pastoral idyll resulting from the unnatural policies of the central authorities in Rome. The society of rustics on the pastoral periphery, where ancient virtues are still upheld, is disrupted by the settlement of foreigners. But Virgil holds forth the promise that these virtues will return in a new golden age. The structure identified in Virgil's mode of discourse can be readily found in "gothicism," a set of ideas, inspired to some extent by Virgil, which first gained prominence during the Protestant Reformation in Germany. German reformers used the term "Goth" to refer to the peripheral Germanic peoples who brought about the destruction of the corrupt Roman empire. Reformers saw themselves as inheritors of this tradition, attempting to overthrow an equally corrupt Roman church and striving to bring about a new religious golden age.[1] In England, in the context of the 17th-century civil war, gothicism received an added political dimension. Parliamentarians identified the Stuart monarchy with governmental traditions inherited from Rome, via the Normans, and their own cause with traditions inherited from the Anglo-Saxon Goths.[2] In Denmark, this in turn was to have implications for the perception of the relationship between Copenhagen, the royal capital, and the Jutland periphery because Jutland, in particular, was identified as the homeland of the Anglo-Saxons. In this way British political debate came to have an influence on the perception of Danish regions.

The first explicit identification of Jutland as a seat of Gothic virtue is in Robert Molesworth's *An account of Denmark as it was in the year 1692.*[3] In this attack on monarchism Molesworth contrasts the former natural state of Denmark with its "unnatural" condition under absolutism since 1664. The book sold widely in Britain and in America where it was a source of inspiration for Thomas Jefferson who traced the ancestry of American republicanism and democracy via English settlers to the Goths.[4] Molesworth was severe in his criticism of absolutism and lavish in his praise of ancient Danish national virtues:

> The Ancient Form of Government here was the same which the *Goths* and *Vandals* established in most, if not all, parts of *Europe*, whither they carried their Conquests, and which in England is retained to this day for the most part. 'Tis said of the *Romans*, that those Provinces which they Conquer'd were amply recompensed, for the loss of their Liberty, by being reduced from their Barbarity to Civility; by the

> Introduction of Arts, Learning, Commerce and Politeness. I know not whether this manner of Arguing hath not more Pomp than Truth in it; but with much greater reason may it be said that all *Europe* was beholden to these People for introducing and restoring a Constitution of Government far excelling all others that we know of in the World. 'Tis to the ancient Inhabitants of these Countries, with other neighbouring Provinces, that we owe the Original of Parliaments. . . .[5]

"Health and Liberty," according to Molesworth, were "the greatest Natural Blessing Mankind is capable of enjoying".[6] Denmark's loss of natural Gothic liberty had consequences for the health of its society. This "womb of nations" had gone barren:

> And although this Country have a tendency to be extreamly populous, the Women being exceedingly fruitful, which is sufficiently proved by the vast Swarms that in former Ages, from these Northern parts, over-ran all *Europe*; yet at present it is but competently Peopled; vexation of Spirit, ill Diet, and Poverty, being great Obstructions to Procreation.[7]

Absolutist policy was repeating the mistake of impérial Rome in favoring foreigners over the native sons of its earth:

> Yet in *Denmark* the Natives are considered much less than Strangers, and are more out of the Road of Preferment, whether it be that the Court can better trust Strangers, whose Fortunes they make, than the Posterity of such whose Fortunes they have ruined: Or whether they think their very Parts and Courage to be diminished in proportion to their Estates and Liberty (which appears to be plainly the case of their Common people) . . .[8]

Molesworth claims to discover the last remnants of ancient virtues preserved on the geographic periphery of Danish society.

> Jutland, [is] part of the ancient *Cimbrica Chersoresus* . . . The Land is more Fertile near the Sea-Coasts; the Inland being full of Heaths, Lakes, and Woods. In short, it is the best Country the King of Denmark is Master of, and appears to be the least declining, because most remote from Copenhagen – Procul a Jove, Procul a Fulmine.[9]

For Danes, Molesworth's book offered a framework of discourse within which the policies of Danish absolute monarchy could be addressed while circumventing the court censorship, and it was widely read.[10]

In Britain and America, discussion of Denmark could serve as an indirect commentary on the English monarchy during the Restoration. In George Puttenham's words, it acted as a "vaile" through which to "insinuate and glaunce" at greater matters.[11]

In the century which followed the publication of Molesworth's account, the gothicist mode of thought helped mould the Danish perception of Jutland and the emerging debate over its future. When the economist and prelate, Erik Pontoppidan, discussed Jutland in his popular *Danske Atlas* of 1763, he presented its present condition in the light of a Gothic past through the device of a quotation from Tacitus:

> "The Cimbrians, who live next to the sea, still occupy the same gulf in Germania. They are a small people today, but their honor is great, and one can still see the broad and wide 'footprints' of their ancient reputation."[12]

Pontoppidan was concerned because Jutland, seat of the heroic Cimbrians, the first tribe to attack Rome, was now so thinly populated. His concerns were reflected in the pages of the influential *Oekonomiske Magazin* which he edited. The magazine represented a departure from absolutist policy because it encouraged public debate on national economic policy. This debate was stimulated by means of essay competitions on topics proposed by the editor.

The journal's first issue of 1757 suggested an essay topic on Jutland's soil fertility:

> Is it possible to cultivate and afforest the various heaths that are lying waste in Jutland, and if so, how would this best be accomplished? What are the true hindrances to this, and how could they best be eliminated without offense to anyone concerned?[13]

The fourth issue turned the debate from the fertility of the soil to fertility of the people:

> Is it reasonable to believe that the population of Denmark and Norway formerly has been greater than it is now? Is it possible, as well as desirable, for the number of inhabitants in Denmark and Norway to become greater than at present? What prevents its growth, and how can these hindrances best be eliminated?[14]

Response to these two topics shows that the reading public related these issues and the heaths themselves with gothicism. Niels Schelde, who did not forget his peasant origins when he rose to the rank of "Royal Counsellor," argued that, if the native farmers received greater "freedom", their numbers would rise "naturally"; this in turn would lead to the intensification of agriculture on the heaths:

Figure 2.1 An illustration from Justus Möser's *Osnabrückische Geschichte*, 1768, showing Gothic family life.

> If mountainous Switzerland can produce so many people that it swarms with them, as if they simply spring out of the mountains . . . should it not be possible for the descendants of the ancient Cimbrians in fruitful Denmark, under as mild a government as any country can pride itself with, to regain together with the increase of its population, the lost courage and bravery of former times?[15]

While Schelde relied upon historical evidence to argue that Jutland had once been more heavily populated, a prize-winning essay on heath cultivation by Søren Testrup used the landscape itself as evidence. Testrup argued that, if the natives of Jutland received freedom, respect, guidance and support, God would

> give his blessing, and richly reward the humble effort and it would be possible to establish the most beautiful sheep nurseries. One could plant trees, and in time the loveliest forests and gardens would spring up. There needs no proof that forests can be planted and grow on the Alheath and other heaths, for stumps and remains of trees found everywhere on the heaths clearly show that in ancient times stately forests grew here.[16]

Just as Virgil's first *Eclogue* contrasted the fruitful, tree-covered landscape of the golden age with the present-day barren landscape with its trees shattered and its shepherds driven away by environmental infertility, so Testrup draws a similar contrast. Not only has the forest been reduced to stumps, the heath itself is

> filled with wolves which hide themselves in the high heather and cannot be driven away . . . The wolves form such large packs at times that the travelers must fear them; they chase the shepherd from his herd, and do great damage.[17]

Despite Testrup's literary licence in exaggerating the threat of wolves (then in fact on the point of extinction) he was correct in claiming that the remains of former tree cover could be found on the heaths, both in the form of stumps and trunks preserved in the bogs and as scrub, twisted by the wind. Like Virgil, Testrup regarded the tree as a symbol of a former fertile time, which could be restored. As described by Testrup, such restoration would follow a natural order: from sheep nursery to garden. The return to fertility is not spontaneous, as Virgil would agree, it requires a commitment by man to participate in the natural process of intensification and increased reciprocity.

Editorship of *Danske Atlas* and responsibility for the volumes specifically on Jutland devolved after Pontoppidan's death in 1764, on his son-in-law, Hans de Hoffman. de Hoffman was a provincial governor in a heath district who was able to combine agricultural insight gained from agronomic training with a sympathetic understanding of the native culture, which derived from a considerable interest in art and cultural history. At a time when the central government was urging greater control over the peasantry in Jutland, de Hoffman argued that agriculture could be improved if the peasantry were given greater freedom and responsibility in the development of native forms of cultivation. He placed great store on the Gothic heritage of Jutland, to which he devoted considerable space in the *Atlas*:

> It is most reasonable to trace the peninsula's name from one of the first peoples who lived there, namely the Guti, Gothi, Giotae, Juti, Juter,

Figure 2.2 A tree trunk found in a bog on the heath (Herning Museum).

> which all mean the same as the still current "Jutlanders". These Jutes, or Jyder, who are the true Goths . . . together with the neighboring Angles and Saxons, are the noteworthy peoples who were called over to Britain in the fifth century to act as support troops, but took the opportunity to prepare the way for more of their countrymen to capture and possess the country. It thereafter received the new name, England. It was presumably this same people who, several hundred years earlier, were considered to be fearsome and respectable warriors by the Romans, and were therefore given the name of Cimbri . . . [18]

It is this Gothic spirit which de Hoffman felt was preserved by the Jutland farmer:

> It must be admitted by everyone, everywhere, that the farmers of Danish Jutland who, from the most ancient times, have lived in greater freedom than most other Danish farmers, are in a much better position, and are of a more diligent and honest sort and nature, than are the farmers of certain other provinces . . .[19]

Virgil had located the preservation of both original national virtues and an ancient artistic heritage in the pastoral periphery. Hans de Hoffman, who was deeply interested in the arts, also sought such a cultural heritage on the grazing landscape of the Jutland heaths:

> No one can deny that many more remains of the true ancient Danish

> language are to be found amongst the common native folk of Jutland than in any of the other provinces. This is especially so deep in the countryside, where contact with foreigners has not been so great, and, as a result, there has been less opportunity either to borrow foreign words or change the character of the language . . . From this it follows that those who would adorn and enrich the Danish language rather than looking always to foreign sources, by which the language is changed more than it is improved, should go now and then to the sources we have here ourselves. By listening to the Jutland mode of speech they would find a supply of pure Danish which would give our language, and especially the bardic art, both wealth and beauty.[20]

Counter-Gothicism

Gothicism applied to the Jutland heaths was explicitly or implicitly critical of government policy. It argued that the freedom of a people increased with distance from the seat of power. If their own fertility or that of their soil left something to be desired, this was because the authorities had managed to overcome distance and constrict their natural freedom. The Crown could not afford an appealing gothicist ideology to form a popular opposition to its own image of enlightened centralism, and so the government commissioned a French historian, Paul Henry Mallet, to write a new history of Denmark – a work which proved to be a subtle defense of the *status quo* within the overall framework of gothicist ideas. Mallet's *Introduction à l'histoire de Dannemarc* was published in 1755. A Danish translation was published in 1756 and 14 years later Bishop Thomas Percy's English translation, *Northern antiquities*, brought Mallet's ideas to Britain,[21] where it had enormous influence as the mainspring of the subsequent vogue for Norse study.[22]

According to Mallet, Rome had itself become "barbarous" through the deleterious effects of slavery. The Scandinavians, however, as the instrument of "nature" put things right once again:

> But Nature had long prepared a remedy for such great evils, in that unsubmitting, unconquerable spirit, with which she had inspired the people of the north; and thus she made amends to the human race, for all the calamities which, in other respects, the inroads of these nations, and the overthrow of the Roman Empire produced.
>
> "The great prerogative of Scandinavia (says the admirable Author of the Spirit of Laws [Montesquieu]) and what ought to recommend its inhabitants beyond every people upon earth, is that they afforded the great resource to the liberty of Europe, that is, to almost all the liberty that is among men. The Goth JORNANDES (adds he) calls

> the north of Europe THE FORGE OF MANKIND. I should rather call it, the forge of those instruments which broke the fetters manufactured in the south. It was there those valiant nations were bred, who left their native climes to destroy tyrants and slaves, and to teach men that nature having made them equal, no reason could be assigned for their becoming dependent, but their mutual happiness."[23]

In these passages "nature" has a dual meaning. Certain aspects of the classical sense of original state remain, but nature has been reified and the sense of developmental process lost. Nature is an original state which determines the character of the nation and, as will be seen, it has been moved back in time from a pastoral stage to a preceding wild or savage stage. The logic by which the classical conception of nature has been transformed is worth examining in some detail.

Like Lucretius, Mallet and Montesquieu see man in the savage state as a product of the wild environment. Lucretius's savage accepts "as a free gift what the sun and showers had given and the earth had produced unsolicited".[24] His self-centered existence, which Lucretius regards as the antithesis of society, is a consequence of this dependence.

Mallet and Montesquieu likewise see the primitive stage of human life as a product of the wild environment, but, unlike Lucretius, they cast such primitivism in a positive light. It is the source of Scandinavian freedom:

> because they inhabited an uncultivated country, rude forests and mountains; and liberty is the sole treasure of an indigent people. . . .[25]

In part, the explanation for this liberty was to be found in natural "science". Forests, they argued, fostered the cool, damp climate which was responsible for both Scandinavia's fertility and its independent spirit. Clearing the forest by contrast, created a warmer and drier climate moderating fertility and favouring human submission to collective restraints. The Scandinavians are described as conceiving their democratic constitution in forest clearings.[26] Felling the forest is linked to civilizing change in manners.[27] According to this mode of reasoning, the ideal environment combines forest with clearings. Through clearance, a people liberates itself from submission to its environment: "A nation is never solely influenced by climate, except in its infancy; while it is uncultivated and barbarous."[28] But total clearance, as in the case of Rome, destroys the necessary remnants of freedom-giving woodland and brings with it slavery.

There is a subtle but vital distinction between the classical conception of nature and that of Mallet and Montesquieu. Nature has ceased to be a normative principle against which social values are measured. Now nature is the physical environment and the source of such values. This in turn can provide a basis for the defence of absolutist *status quo*. It is a good example of the way in which the use of the word nature can, in Lovejoy's words, "slip more or less insensibly from one ethical or esthetic standard to its very antithesis, while nominally professing the same principles."[29]

Mallet's argument challenges the assumptions of the gothicists, while employing much of their rhetoric. For them, the essential nature of the nation was already determined in its natal state. Consequently, Mallet argued, the physical and social character of the nation could be preserved even under a non-gothic form of government:

> Is it not well known that the most flourishing and celebrated states in Europe owe originally to the northern nations, whatever liberty they now enjoy, either in their constitution, or in the spirit of their government? For although the Gothic form of government has been almost everywhere altered or abolished, have we not retained, in most things, the opinions, the customs, the manners which that government had a tendency to produce? Is not this, in fact, the principal source of that courage, of that aversion to slavery, of that empire of honour which characterise in general the European nations; and of that moderation, of that easiness of access, which so happily distinguish our sovereigns from the inaccessible and superb tyrants of Asia?[30]

Gothicism and regional development

Gothicists like de Hoffman had argued that the present condition of the heaths, with their low fertility and lack of population, resulted from the disruption of the natural society of the native inhabitants, implying criticism of the responsible external authority.[31] Mallet's "counter-gothicism," on the other hand, suggested that it was the heath environment which formed the character of its native inhabitants, not the government. This conception of the relationship of people to their environment would in turn have consequences for development policy in the region.

The central authorities ignored the advice of native experts, such as Hans de Hoffman, who favoured intensification on the basis of traditional local agricultural practice. Instead, they brought in outside expertise, with little sympathy for the native inhabitants and their traditional forms of land use. Native agriculture was based upon extensive grazing on the heath with settlement confined to small hamlets and isolated farmsteads spread along the meadow lands adjacent to streams. This system was in ecological balance because the livestock (sheep, cattle, horses) provided fertilizer to maintain intensive grain cultivation adjacent to the farmstead. This was supplemented by occasional burning and cultivation of the outfield grazing lands. This practice not only brought in extra crops of grain, but it was also a means of improving the pasture because grass, followed by a fresh crop of heather, grew on the fields once cultivation was abandoned. The sandy podzolized condition of the soil worked against intensive grain cultivation, as did distance from markets and bad roads. Such conditions fostered the breeding of livestock (which could walk the long distance to markets in Germany) as well as a welter of secondary occupations such as knitting, pottery making and illegal distilling. Rampant smuggling meant that most of these activities were difficult to control and to tax, so the heath peasantry was unpopular with the central government which, anyway, regarded their agriculture as a wasteful use of the land.[32]

Instead of taking measures to encourage the intensification of local agriculture, for example by making improvements in the transport infrastructure, or by introducing more suitable crops like the potato, which could be incorporated into the existing agricultural structure as a fodder crop, the government attempted to circumvent the native populace. A prominent German economist, Johan Heinrich Gottlob von Justi, was invited to evaluate the developmental potential of the heaths. He proposed a plan for settling 10 000 colonists in large, geometrically planned villages as a method for developing the heath. In the course of the 1760s some 300 families from the fertile wheat-growing district of Pfalz in Germany were settled in villages of between 40 and 60 households. The theory was that, if people whose nature had been formed in a densely populated, intensively cultivated region were brought in, they would reproduce the same environmental conditions on the heath. Johan von Justi, who influenced Johann Heinrich von Thünen, treated the heath as an isotropic plane, and took little consideration of local ecological differences. The villages were often located far from meadow and stream, lacking water supplies and on open terrain which would have been well suited to intensive grain cultivation in more fertile districts.[33] This policy necessitated the displacement of the local inhabitants from the land where the colonists settled, and thereby a reduction in the native resource base.

The von Justi scheme proved disastrous, and the majority of the colonists had left within a decade. The 60 families who remained survived by monoculture of potatoes which provided a saleable crop and their primary means of subsistence – earning them the epithet "potato Germans".[34] Reports from the two surviving villages on the Alheath describe "Irish" conditions with a rapidly expanding population and a dwindling subsistence base. Had von Justi succeeded in bringing in and settling his 10 000 colonists, the potato blight which began to plague Europe shortly thereafter might have wreaked havoc in Jutland.[35]

As a loyal civil servant, de Hoffman agreed to administer von Justi's plan, and was able to ameliorate the situation by decentralizing colony settlement and bringing natives into the scheme once the Germans abandoned the area. Through economic incentives and crop improvements (for example, by introducing the potato into the traditional native crop association) he was eventually able to demonstrate the value of his original ideas concerning the intensification of native agriculture, within the framework of the von Justi scheme.

In de Hoffman's administrative actions the practical significance of gothicist ideas can be seen converging with a pragmatic response to regional development. Hans de Hoffman's gothicism, as promulgated in the *Danske Atlas*, provided the ideological justification for policies of vital importance to the future of Jutland. His efforts on behalf of the native Jutlander came at a time of increasing national sentiment and corresponding dissatisfaction with foreign influence upon national affairs. Through his writing and practical work he was able to give the native Jutlander an important role in this awakening of national identity.[36]

Notes

1 Kliger (1947), The Gothic Revival and the German Translatio, pp. 73–103.
2 Kliger (1952), *The Goths in England.*
3 Molesworth (1694), *An account of Denmark as it was in the year 1692*, 3rd edn. On the history of the book, see Brach (1879), *Om Molesworths Skrift, "An Account of Denmark as it was in the Year 1692"*. The *Account* reached its fifth edition by 1696. It was translated into a number of foreign languages, including French, German and Dutch.
4 Kliger (1952), p. 110.
5 Molesworth (1694), pp. 38–9.
6 Ibid., Preface.
7 Ibid., p. 81.
8 Ibid., p. 75.
9 Ibid., pp. 28–9.
10 Brach (1879) details the public discussion.
11 See Chapter 1, n. 14.
12 Pontoppidan (1763), *Den Danske Atlas*, vol. I, pp. 22–3. The quotation is from Tacitus, *Germania*, XXXVI.
13 Pontoppidan (1757), Introduction, *Oeconomiske Magazin* I, p. xiii. A useful presentation of Danish economic thought at this time is Bisgaard (1902), *Den Danske National Økonomi i det 18. Aahundrede.*
14 Pontoppidan (1761), Introduction, *Oeconomiske Magazin* IV, p. xii.
15 Schelde (1761), Frie=Tanker om Aarsagerne til Folkemangel i Danmark, p. 48.
16 Testrup (1759), Forslag om Hederne i Nørre=Jylland til Ager og Eng at optage, p. 98.
17 Ibid., p. 99.
18 de Hoffman (1768), *Den Danske Atlas*, vol. IV, pp. 5–6. S. Kliger regards the etymological link between "Goth" and "Jute" as "a semantic confusion" in Kliger (1945), The "Goths" in England: an introduction to the Gothic vogue in 18th-century aesthetic discussion, p. 110.
19 de Hoffman (1768), p. 45.
20 Ibid., p. 9.
21 Mallet (1770), *Northern antiquities: or a description of the manners, customs, religion and laws of the ancient Danes and other northern nations including those of our own Saxon ancestors*, 2 vols (T. Percy, anon. trans.).

22 Farley (1903), *Scandinavian influences in the English Romantic movement*, pp. 30–2.
23 Mallet (1770), p. liv.
24 See Chapter 1, n. 22.
25 Mallet (1770), p. 163.
26 Mallet (1756), *Indledning udi Danmarks Riges Historie* (trans. anon.), p. vii: this is the Danish translation of *Introduction a l'Histoire de Dannemarc*. Mallet quotes Montesquieu on this in Mallet (1770), pp. 123–34.
27 Mallet (1770), pp. 410–14.
28 Ibid., pp. 408–9.
29 Lovejoy (1927), "Nature" as aesthetic norm, p. 444. The shift in meaning which occurred in the use of the word nature in Danish literature during this period is elucidated in a seminal study by Bredsdorff (1975), *Digternes Natur*.
30 Mallet (1770), p. liii. On the ideological use of the landscape of nature at this time, see Feingold (1978), *Nature and society*; and Turner (1979). *The politics of landscape*.
31 For de Hoffman's views see: de Hoffman (1757), *Oeconomiske betragtninger om Aarhus= Stift*; de Hoffman (1758), Om Heederne i Jylland. Heedens Dyrknings Forbedring, pp. 20–38, 39–46; de Hoffman (1781), *Samtale Angaaende Hedernes Dyrkning og Forbedring I Jylland, som et Anhang til Hr. Conferentz=Raad og Amtmand Fleischers Agerdyrknings Cathechismus Hvilken det Kongelige Landhuusholdnings Selskab har ladet trykke og uddeele*.
32 On traditional heathland settlement patterns, see Hansen (1971), Rural settlement on Danish glacial outwash Plains, pp. 206–18; Matthiessen (1939), *Den Sorte Jyde*; and Skrubbeltrang (1966), *Det Indvundne Danmark*.
33 For von Justi's plan, see von Justi (1758 [1760]), Allerunterthänigstes Gutachten, Wegen Anbauung der jütländischen Heiden. Keine Oberfläche der Erden ist Schlechterdings und vor sich selbst unfruchtbar, pp. 609–72, 672–9. On the economic theory of von Justi, see Small (n.d.), *The cameralists: the pioneers of German social polity* (orig. 1909). On von Justi as a predecessor of Johann Heinrich von Thünen, see Henning (1972), Die grosse Stadt in verschiedeen Verhältnissen betrachtet (J. H. G. v. Justi, 1764), pp. 186–97.
34 On the colonization scheme, see Lund (1975), Kartoffeltyskerne, pp. 31–66; and Thorpe (1957), A special case of heath reclamation in the Alheden district of Jutland, 1700–1955, pp. 87–121. The latter paper is dependent upon the work of local historians descended from the German colonists.
35 Carstens (1839), *Bemærkninger over Alheden og dens Colonier*.
36 For further information on de Hoffman and the colonization scheme, see: Andersen (1970), *Den Jyske Hedekolonisation*; Rasmussen and Lund (1974), *Hedekolonierne*; and Aakjær (1919a), Heden: in *Samlede Værker*. vol. V, pp. 599–710.

3 *Nature as wilderness landscape: Ossian's and Blicher's heath*

The author Steen Steensen Blicher (1782–1848) was a central figure in determining the 19th-century perception of the Jutland heath. He created an image of the heath as wilderness and proclaimed that wilderness was a repository of natural virtues. At the same time he was able to question this perception of the heath and of wilderness, returning to a conception of nature much closer to the classical, but one which stressed the particularity of place rather than cosmological principles.[1] Blicher's ideas on the heath were inspired in part by Danish gothicists, but they were also rooted in ideas then emerging from another European peripheral region: Scotland.[2]

One of the places where Montesquieu's ideas resonated most deeply was mid-18th-century Scotland. Intellectuals like Henry Home (Lord Kames), author of *Elements of criticism* and *Sketches of the history of man*, Adam Smith, the economist, author of *Considerations concerning the first formation of languages*, and Hugh Blair, author of *Lectures on rhetoric and belles lettres* met in Edinburgh to engage in "conjectural" historical enquiry concerning the primitive origins of art, language and society. Their ideas were soon given empirical warranty in 1759 when a young protégé of the group, James Macpherson, returned from the Highlands with a body of verse, and later epic poetry, supposedly composed by an ancient Gaelic bard named Ossian. It had been, he claimed, preserved by oral tradition, and translated from the *Erse* by Macpherson himself. The poetry so perfectly corroborated the conjectures of the Edinburgh group on the ideal qualities of life in the wild that no need was felt to check the veracity of the Gaelic original. This soon "disappeared".[3] Despite suspicions, Macpherson's ability to combine fragments of actual Scottish folklore within an epic framework constructed according to ideas of his mentors made it difficult to prove that he was the effective author of the work. Some patriotic Scot could always be found to testify that a given passage was genuine.[4] Several volumes of epic poetry were eventually published, complete with preface by the influential aesthetician, Hugh Blair, testifying to their historical origin and significance.[5]

A popular vogue for Ossian's works spread throughout Europe in the last years of the 18th century, culminating in Ossian's near deification by Napoleon. In Germany followers of Friedrich Klopstock, known as "the bards," went so far as to develop a bardic life-style, which required the poet to compose while sitting under an aged oak, harp in hand, head wreathed in oak rather than laurel.[6] As a frequent guest, Klopstock himself brought the enthusiasm for Ossian to the Danish court. However, those Danes who did not read English had to wait until 1807 for a young Jutland poet named Steen Steensen Blicher to publish a full Danish translation.[7]

The poetry of Ossian held an immediate appeal for Blicher. Only later did he become aware of and begin to question the social implications of Ossianic aesthetics. Ossian's poetry like the work of Mallet had much in common with aspects of the gothicist tradition. Blicher was aware of this tradition from his contacts with Hans de Hoffman, who had been a guest in his childhood home at Vium vicarage and his *Danske Atlas* had a prominent place on the family bookshelf. Blicher's father, Niels Blicher, was himself the author of a topographical work on Vium parish, which bordered the vast Alheath, and, like de Hoffman, he had defended the native Jutland economy and culture in the face of foreign colonization.[8] The newly discovered "ancient" Ossian poetry corroborated gothicist ideas about the cultural importance of such isolated peripheral regions. Its glorification of a heath-covered landscape lent, by implication, new status to Jutland. Macpherson claimed of his Scots, as de Hoffman had written of his Jutlanders, that "their language is pure and original, and their manners are those of an ancient and unmixed race of men". They "differed materially" from those who possess the "more fertile part of the kingdom". Macpherson even argued that the ancient Gaelic peoples were related to the Goths and Jutland figures in Ossian poetry.[9] The Ossianic works lent support to the idea that marginal heathlands were culture reservoirs of an ancient civilization, much as the heath landscape, with its barrows and tree stumps, preserved remnants of a more glorious past. It is from this landscape, from its hills and mists and blasted oaks, that Ossian drew inspiration for his poetry.

Like Macpherson, Blicher was well educated in the classics and would have found the Ossian poetry familiar not only because of its agreement with gothicist precepts, but also in its resemblance to Virgil. Even Blair had commented upon the extraordinary resemblance of Ossian's "elegiac strain" to Virgil's.[10] Both recall with wistful nostalgia a lost, tree-clad golden age.[11] But in the Ossianic landscape Virgil's pastoral scene is given a new, more wild energy. For Blair it is sublime rather than beautiful:

> The events recorded are all serious and grave; the scenery throughout, wild and romantic. The extended heaths by the seashore; the mountains shaded with mist; the torrent rushing through a solitary valley; the scattered oaks; and the tombs of warriors overgrown with moss; all produce a solemn attention in the mind, and prepare it for great and extraordinary events.[12]

Figure 3.1 Ossian, painted about 1785 by the Danish painter Nicolai Abraham Abildgaard (State Museum of Art, Copenhagen).

Blicher found the elegiac strain congenial. It runs through his own work, and he readily identified with the figure of the ancient bard. As in the case of classical Arcadians, song played a vital role in creating the Goths' ollective identity because it was a means of recalling their past. According to the *Danske Atlas* "every king had one or more so-called bards," whose task, Macpherson tells us, "consisted in hearing or repeating their songs and traditions, and these entirely turned on the antiquity of their nation, and the exploits of their forefathers".[13] The young Blicher adopted the Ossianic bardic posture as, for example, in the 1817 poem *Jutland journey in six days.*

> Lonely I lay on my heather overgrown hill,
> The roar of the storms passed over me,
> The mossy gravestone was under my neck,
> Up in the clouds lingered my gaze.[14]

Blicher could not have found a better theme. His verses soon established him as the Jutland bard and a foremost national romantic poet.

> Hail, you my land of birth, our forefather's dark
> brown burial place;
> With your barrows, the heather-thatched
> houses of the dead.
> Here rust the swords, from which the world
> has shivered;
> Here crumbles that strength, of which only
> legend remains.[15]

Together with writers like B. S. Ingemann, N. F. S. Grundtvig and Adam Oehlenschläger, Blicher did much to raise the self-image of a recently defeated nation. In 1814 Denmark had lost Norway to the Swedes and faced economic bankruptcy. By reviving the memory of past glories these writers created a golden age in art, if not in reality, and through Blicher's efforts the heaths became established as that landscape which preserved, under its peat and moss, the heroic character of the nation.[16]

Second thoughts on Ossian

As its social implications became apparent to him, Blicher began to revise and question the Ossianic posture which he had adopted. This may have been due, initially, to B. S. Ingemann's 1815 review of the first volume of his poetry. Ingemann, who was a prominent historical novelist, commented disapprovingly upon Blicher's faithful pursuit of Ossian onto the "bloody heath," and asked why he followed this "cheerless spirit" into its grey twilight when the sun had long since risen over a more civilized world.[17] Blicher apparently took the point, even if he did not fully abandon the bardic pose. In *Jutland journey in six days* (1817) Blicher opens in an Ossianic mood, but then poses a rhetorical question:

> Why do you stare, eye:
> At the shadows of ancient times? lives your

Figure 3.2 The heath in an Ossian fog.

> happiness alone
> In departed years?[18]

Why, he continues, should a bard not praise such recent events as the efforts of the people of the Alheath to restore their lands to their original state, "dressed in smiling green," with "fragrant meadows, grain rich fields and sheltered forests":[19]

> Why sings the bard so gladly of war, and tempt
> the hearts of the powerful with lays of
> bloody laurels?
> . . .
> Are the Alheath's heather-thatched houses
> Not just as beautiful as the castle's smoking
> foundations?
> Shall thistles and thorns delight the
> leaders' hearts
> More than rich grainfields?[20]

These questions strike at the foundation of the Ossianic aesthetic, the glorification of the most primitive stage in human development. According to Blair:

> There are four great stages through which men successively pass in the progress of society. The first and earliest is the life of hunters; pasturage succeeds to this, as the ideas of property begin to take root; next agriculture; and lastly commerce.

Ossian, in Blair's view, belonged to the first stage, though "pasturage was not indeed wholly unknown".[21] There are traces of the classical conception of nature, but in Ossian normative social values derive directly from the wild state. Macpherson believed that the first stage was "formed on nature," and hence the most "noble" while Blair felt that Ossian spoke with "the powerful and ever-pleasing, voice of nature":[22]

> It is a great error to imagine that Poetry and Music are Arts which belong only to polished nations. They have their foundation in the nature of man . . . In order to explore the rise of Poetry, we must have recourse to the deserts and the wilds; we must go back to the age of hunters and of shepherds; to the highest antiquity; and to the simplest form of manners among mankind.[23]

Ossian's poetry glorified precisely the landscape of "thistles and thorns," and not that of "rich grainfields". As Blair tells us: "everywhere, the same face of rude nature appears; a country wholly uncultivated, thinly inhabited, and recently peopled. The grass of the rock, the flower of the heath, the thistle with its bur, are the chief ornaments of his [Ossian's] landscape. 'The desert,' says Fingal, 'is enough to me, with all its woods and deer.' "[24] Ossian did not glorify the emergence of agriculture. On the contrary, he bemoaned the loss of the heroic society which preceded it, seeing it replaced by a race of weak men.

Despite superficial resemblances, the Ossianic aesthetic ran counter to gothicist ideals. The gothicists saw development towards more intensive forms of land use as natural and the heath as an unnatural state. The Ossianic aesthetic, by contrast, regarded development as unnatural; wild untouched landscape was therefore the most natural. As a poet and as an avid hunter, Blicher might have liked to escape into "Ossian's heath" but in practice he actively engaged in and supported its cultivation. As a pastor in a heath district his very livelihood depended upon its cultivation. His question, "shall thistles and thorns delight the leaders' hearts more than rich grainfields?", was more than rhetorical.

Blicher was disturbed by the parallels which he observed in social and economic conditions between peripheral areas in Britain and Jutland. He was a keen Anglophile, well aware of the political and social polarization developing in the wake of industrialization. He admired Oliver Goldsmith's critique of social change and translated *The vicar of Wakefield* into Danish. Goldsmith, of course, was known for his attack on aesthetic glorification of landscapes emptied of people:

> The man of wealth and pride,
> Takes up a space that many poor supplied;
> . . .
> His seat, where solitary sports are seen,
> Indignant spurns the cottage from the green.
> . . .

> Thus fares the land, by luxury betray'd
> In nature's simplest charms at first arrayed,
> But verging to decline, its splendours rise,
> Its vistas strike, its palaces surprise;
> While scourged by famine from the smiling land,
> The mournful peasant leads his humble band.[25]

Blicher also admired Sir Walter Scott, calling him "the historian of humanity". He contrasted Scott's self-effacing concern for the social collectivity of rural Scotland with what he regarded as the destructive egotistic individualism of Lord Byron.[26] Scott himself, like Blicher, feared for the condition of the people in rural districts and likewise was engaged in land improvement.[27] Like Blicher, Scott had matured from a youthful infatuation with Ossian to question the glorification of empty wild landscapes, not least because it legitimized the depopulation of the Scottish hill farms to make room for hunting reserves, forests and sheep:

> in but too many instances, the glens of the Highlands have been drained, not of their superfluity of population, but of the whole mass of the inhabitants, dispossessed by an unrelenting avarice, which will one day be found to have been as unwise as it is unjust and selfish. Meanwhile the Highlands may become the fairy ground for romance and poetry: the subject of experiment for professors of speculation, political and economical.[28]

Blicher also criticized the anti-social glorification of wild landscape. He applied this critique, furthermore, to his own image as a lover of the wild heath. This can be seen in the introductory landscape depiction in his classic short story, "The hosier" (1829). This *tour de force* is ironic and paradoxical, reflecting the author's own ambivalence to his role as the lonely poet of the heath.[29] In the opening passage the narrator, a romantic wanderer easily recognizable as Blicher himself, is identified with the wild desert nomad in his disdain of permanent agrarian settlement.[30]

> Sometimes when I have wandered across the great heaths with nothing but brown heather about me and blue sky above me; when I have strolled far from human beings and the marks of their puttering here below – mere molehills that time or some restless Tamburlaine will level to the ground; when I have flitted, light of heart, proud of my freedom like the Bedouin whom no house, no narrowly bounded field ties to one spot, who possesses all that he sees, who lives nowhere, but roams as he pleases; when in such a mood my roving eye has caught sight of a house on the horizon unpleasantly arresting its airy flight, then have I wished – God forgive me the passing thought, for it was nothing more – would that this human dwelling

> were not there! For it harbors trouble and pain; there people quarrel and wrangle about mine and thine.[31]

Like the restless and destructive nomad, Tamburlaine, the wanderer would level settlements to the ground, though he acknowledges that the act would run counter to prevailing morality. He protests that this is only an impulse and that he did not entertain it, but his protest is not entirely convincing for already in the next paragraph the idea returns to him:

> A forester has proposed that the entire colony development be wiped out, and that trees be planted in the fields and on the site of the razed hamlets. I have sometimes been seized by a far more inhuman idea. What if we still had the heather-grown heath, the same that existed thousands of years ago, undisturbed, its sod unturned by human hands! But, I repeat. I did not mean it seriously. For, when exhausted, weary, languishing with heat and thirst, I have longed intensely for the Arab's hut and his coffeepot, then I have thanked God for a heather-thatched cottage – though miles distant – promising me shade and refreshment.[32]

In this second paragraph it is no longer a nomad but a forester who would level human settlement. The destructive force, emptying the landscape of cultivators, is no longer embodied in the primitive, but in a figure that Scott might term a "professor of speculation". Both have in common the desire to return the heaths to a more extensive and extractive form of land use requiring vast areas of land unencumbered by the boundaries imposed by mixed farming. In this they oppose the gothicist's ideal landscape of arable clearances and meadow glades sheltered by surrounding forest. The nomadic herder desires an open grazing landscape of heath, while the forester would restore it to a wooded state, a state believed by the gothicists to have preceded the heath, and identified with the figure of the hunter.[33] The romantic wanderer needs to "possess all that he sees" in order to find esthetic pleasure in the landscape.

> And I was in just such a state one calm, hot September day, some years ago, when I had walked far out on that same heath which, in an Arabian sense, I call my own. Not a breath of wind stirred the reddening heather; the air was sultry and drowsy. The distant hills that bounded the horizon swam like clouds circling the immense plain, and took on the forms of houses, towers, castles, human beings, and animals, but all in dim, formless outlines, wavering and unstable like pictures in a dream. One moment a hut was transformed into a church and the church into a pyramid; there a spire shot up, and there another sank down; a man became a horse, and the horse an elephant;

> here rocked a boat and there a ship with all sails set.
>
> My eye feasted for a long time in contemplation of these fantastic images – a panorama such as only the sailor and the desert-dweller have an opportunity to enjoy. But presently, feeling tired and thirsty, I began to search for a real house among the many false; I earnestly desired to exchange all my magnificent fairy palaces for a single human cottage.[34]

The figure of the romantic wanderer in "The hosier" is characterized by the self-irony with which Blicher endowed his literary alter egos. In his ability to create a fairy ground from the landscape he forgets more basic, earthly needs for food and human companionship. This conflict between a desire for personal pleasure, either esthetic or material, and collective human needs, is the main theme of the story. It contrasts a miserly hosier's concern for economic gain with his daughter's need for human warmth and love. The narrator sides initially with the hosier, but when the tragic consequences of his cold neglect of human needs become apparent, the narrator revised his opinion, rather as he had revised his image of a fairyland heath when tired and thirsty. The story conveys Blicher's conflict between the escapist dreams of his poetry and the more practical material needs of the heathland natives of which he was one.[35] In "The poet's bliss," a poem written in the same year as "The hosier," Blicher writes:

> Though his way was narrow and dark,
> He made a paradise of a desert
> He swims in a sea of ideals –
> (quietly) he dreams.[36]

Blicher eventually found a personal solution to his dilemma. In later work he largely abandons Ossianic landscape depiction for empathetic characterization of life and culture among the common agriculturalists of Jutland. He did not limit himself to this regional theme, however; he also wrote stories concerned more generally with social issues and psychological conflict. In this he may be regarded as one of the first realists in Danish literature.

Natural science and wild landscape

The Ossianic interpretation of wild nature and hunters possessing primeval nobility is reflected in part by Blicher's interest in the geology of Jutland. An appreciaton of the natural sciences is present throughout Blicher's landscape description. In his introduction to "The robber's den," he speculates upon the origin of the landscape:

Figure 3.3 A house on the Alheath, painted in 1912 by Hans Smidth (State Museum of Art, Copenhagen).

> This high ridge of land one imagines, not without reason, to have been the first part of the peninsula to come to view, rising from the sea and tumbling it to both sides, where the rolling waves washed together hills, and hollowed out valleys.

The cultural landscape, however, remains his preoccupation:

> Narrow roads with deep ruts separated by high ridges indicate less travel and less intercourse between the inhabitants. The houses of the people become poorer and poorer, lower and lower, the farther we go, as if they were bowing to the violent onslaught of the west wind. As the heaths become larger and more frequent, the churches and villages become fewer and farther apart. On the farms the light frames for drying hay give way to stacks of black peat and the orchards to cabbage plots.[37]

In Ossian's poetry the emphasis had been upon wild, untouched landscape with its endless vistas. Macpherson saw wild landscape as a ruin created by physical processes out of a pre-existing harmonious landscape. The elegiac strain in his work thus loses human reference. "The globe," he mourns, "though smoothed over into some regularity by the flux of time, appears not to be what it once had been."[38] This elegiac geology owes its origin largely to Thomas Burnet's geological study, *The sacred theory of the earth* of 1684. This work by a member of the (British) Royal Society, inspired writers on landscape well into the romantic era. For Burnet, mountains

Figure 3.4 A road on the Alheath.

were the landscape "ruins" of an antediluvian golden age. Blending classical and Biblical sources, he argued that:

> The greatest objects of Nature are, methinks, the most pleasing to behold, and next to the great Concave of the Heavens, and those boundless Regions where the stars inhabit there is nothing that I look upon with more Pleasure than the wide Sea and the Mountains of the Earth. There is something august and stately in the Air of these Things, that inspires the Mind with great Thoughts and Passions; We do naturally, upon such Occasions, think of God and His Greatness: and whatsoever hath but the Shadow and Appearance of INFINITE, as all Things have that are too big for our Comprehension, they fill and over-bear the Mind with their Excess, and cast it into a pleasing kind of Stupor and Admiration.
>
> And yet these Mountains we are speaking of, to confess the Truth, are nothing but great Ruines; but such as show a certain magnificence in Nature; as from old Temples and broken Amphitheaters of the *Romans* we collect the greatness of that People.[39]

Burnet was an exception. Prior to the 18th century, mountains and wilderness were seen by most people as unnatural blemishes upon the earth; a view, of course, that harmonizes with the classical conception of nature in identifying such environments with the savage, unnatural state of man. Works carrying the mantle of natural science, not least Burnet's, provided a basis for changing attitudes to wild landscapes.[40] Burnet transposed to wilderness older notions of nature as process with the consequence that ideas and feelings earlier associated with the development of human society are transferred to material nature. In the tradition of elegy we find wild landscape identified with a Roman ruin, ultimately with the ruin of the golden age landscape itself. But the ruin is no longer primarily a consequence of human action, but of physical processes. Likewise the esthetic impact of the ruin is no longer identified with social implications; rather, the impact of greatness in size and distance is the source of esthetic satisfaction. The mind is "naturally" inspired to great thoughts and passions by grand scenes from "nature". It is this esthetic that was identified with notions of the sublime formulated by Blair and his protégé Macpherson among others. For them, man in the wild state, untouched by civilization, is most responsive to the grandeur of space created by mountains and bare heaths. In the poetry of such men the sublime awe created by these scenes is directly recorded.[41] It is against this background that we can measure the significance of Blicher's emphasis upon the heath as a cultural landscape, formed by men, and signifying their deeds. Similarly, it is against the background of cultural and economic change being experienced in the Highlands that we must evaluate Blicher's rejection of the Ossian esthetic.

Social and economic change and the taste for wilderness

The Renaissance scientific revolution went hand-in-hand with the exploration and conquest of distant parts of the world for trade and resource exploitation. Advances in the natural sciences made possible the developments in cartography and navigation so necessary for long voyages to distant and virtually unknown parts of the world. The figure spying out over a vast untamed landscape need not be a natural scientist; he could equally have been an explorer conquering a distant, wild continent. Just as Burnet grounded his theology in observations of material nature, and Blair and Macpherson grounded their esthetics in the psychic impact of the landscape, so too the conquest of the globe's material nature led, as Basil Willey has argued, to new normative conceptions of human behavior:

> The centuries following the Renaissance liberated the acquisitive impulses, also in the name of Nature, and severed economic ethics from control by any comprehensive conception of the ultimate purpose of human (not to say Christian) living. For, as the eighteenth century discovered,

> Thus God and Nature fixed the general frame,
> And bade self-love and social be the same.
>
> . . . the new world offers us as its symbolical figure Robinson Crusoe, the isolated economic man, pitting his lonely strength successfully against Nature in a remote part of the earth . . . The Law of Nature, which in the Middle Ages had been a check on unregenerate impulse, had now been transformed into a sanction for *laissez-faire* and free competition for the spoils of the world.
>
> Somewhat similarly the idea of a State of Nature, especially after Locke, came to be used as a means whereby the new ruling classes could vindicate, against the surviving restraints of the old feudal and ecclesiastical order, their cherished rights of individual freedoms of property.[42]

The savage individual, disdained by Lucretius, having no thought of the common good, living and fending for himself, carrying off on his own account such prey as fortune brought him, now became a social ideal.[43] The change was not, however, immediate. Prior to the 18th century "the acquisitive urges still hid behind religious and political outworks," but behind the rhetoric of ideological sanction it is clear that "what both individuals and states were really after were the material spoils of the world, now lying readier for exploitation than they had ever been for the Tamburlaines".[44]

Willey reminds us of the material and historical context within which the conception of nature symbolized by figures like Crusoe, Ossian or Tamburlaine emerged; we are dealing with more than literary metaphor, more even than ideology – with the transformation of Europe's economy, and the concrete human consequences thereof. The Scottish Ossian provides a commentary on the realities of 18th-century Highland society and its change. Macpherson's bard bemoans the coming of a new unheroic age of weak men. The society of hunters and primitive pastoralists, which he glorifies, existed prior to that of the crofters whose agriculture formed the economic backbone of contemporary clan society. In classical terms it would have been most unnatural to idealize a wild society because the natural process of human development was towards increasingly intensive and less extractive uses of the land. A reversion to pastoralism would be equally unnatural because in Varro's words, "grazing cattle do not produce what grows on the land, but tear it off with their teeth".[45] But it was just such a society that Macpherson idealizes at a time when agriculturalists were being evicted from the Highlands and "sheep ate men". After the defeat of the clans in 1745, property ownership by individual lairds replaced collective clan control of land under the rule of the chief. Under the new regime the estate owner, frequently a former clan chief, was no longer bound by the inherited clan customs of economic and social reciprocity. Consequently, many took the opportunity to clear the land of crofters and turn it over to sheep, forestry, or hunting preserves for men of wealth from the south.[46] Such unnatural reversion in classical terms was legitimized, or naturalized, in the Ossian esthetic. The restoration of land to a former wild state reminiscent of the heroic Ossianic age no doubt had appeal to those who were engaged in breaking down clan society. It would be an oversimplification to suggest that Macpherson constructed the Ossian works directly to underwrite the 1785 clearance of his own Highland estate, Belleville – one of the earliest of such acts. However, the ideological value of the Ossian body of work to those who promoted and benefited from such action is too obvious to ignore.[47]

The Ossian esthetic effectively denies its economic and social implications through its claim that the poetry, indeed the human context itself, in which Ossian was claimed to have lived, were offered as products of wild physical nature, which, by definition, was pre-social. Not only were Ossian and his poetry products of wild nature, but poetry itself, and with it the esthetic pleasure to which it responded, were seen to spring from the direct experience of wilderness. For this reason it was critical that Macpherson hide his own authorship of the works; to admit it would risk losing most of its impact. The poetry pretends to be inspired by wild nature, not by social and economic realms of discourse. The intellectual circles that produced the Ossian esthetic on the other hand, were permeated by such discourse. A key figure in these circles was Adam Smith who began his career as an esthetician, preceding in this role his protégé, Hugh Blair, at the University of Edinburgh.[48] The connection between esthetics and economics may not be strong in modern intellectual discussion, but it was readily apparent to Karl Marx a generation later. Referring to Smith's social assumptions, he pointed out that

> The individual and isolated hunter and fisherman, with whom Smith and Ricardo begin . . . in no way express merely a reaction against over-sophistication and a return to misunderstood natural life, as cultural historians imagine . . . This is the semblance, the merely aesthetic semblance . . . It is rather, the anticipation of "civil society," in preparation since the sixteenth century and making giant strides towards maturity in the eighteenth. In this society of free competition, the individual appears detached from the natural bonds etc. which in earlier historical periods make him the accessory of a definite and limited human conglomerate. Smith and Ricardo still stand with both feet on the shoulders of the eighteenth-century prophets, in whose imaginations this eighteenth-century individual . . . appears as an ideal, whose existence they project into the past. Not as a historic result but as history's point of departure. As the Natural Individual appropriate to their notion of human nature, not arising historically but posited by nature.[49]

To highlight Macpherson's deception and Marx's critique is not to imply conscious deceit on the part of the conjectural historians, but rather to bring to light a false consciousness implicit in the whole framework of discourse, wherein the historical relationship between nature as landscape object and nature as developmental norm remains unexamined.[50]

In this context Blicher's importance lies in the transition he succeeded in making from acceptance of the Ossianic ideology to critical examination of its social and economic implications. This led him in later work to focus upon the development of society rather than to dwell on the depiction of scenery. For Blicher, as for classical writers, the state of development of the environment provided a metaphor for the nature of society, the nature of society did not derive from its environment.

Blicher's writing carefully mediates the tension between an Ossianic conception of nature and an older, more classical conception. Unlike Macpherson, Blicher did not reify nature as physical object and thus displace *social* values with *individualist* values, derived directly from the material environment itself. Blicher, however, does differ from the classical writers in so far as he is concerned with the material character of a specific locality, as well as with the grander theme of the general development of society and the nation. He thereby mediates the Ossianic view of nature as thing and the classical conception of it as a cosmological principle.[51] Nature is transferred from the cosmos to the material specificity of particular places – in this case Jutland. In his concern for the region of his youth, the field of care to which he dedicated his life, Blicher was a product of his age. As Arthur O. Lovejoy stresses, the emphasis upon diversity and particularity is the one common factor in the otherwise diverse tendencies which have been termed "romantic":

> the quest for local color; the endeavor to reconstruct in imagination the distinctive inner life of peoples remote in time or space or in cultural condition; the *étalage du moi*; the demand for particularized fidelity in landscape-description, . . . the cultivation of individual national, and racial peculiarities.[52]

Blicher's characters are not posited by nature, they arise historically while remaining deeply implicated in their material environment. Unlike Virgil's Arcadia, Blicher's Jutland was a real place, a place one could visit. One could find the gravestones of his heroes in the cemetery by the church where he preached. One could wander thence out on to the heath, as so many people did, inspired by Blicher's work. The interest in the cultural landscape of the Jutland heath which Blicher created was to have an enduring effect on both the perception and later use of the Jutland environment.

Notes

1 Olwig (1974), Place, society and the individual in the authorship of St. St. Blicher, pp. 69–114.
2 Key works on Blicher for this study are: Baggesen (1965), *Den Blicherske Novelle*; Nørvig (1943), *Steen Steensen Blicher, Hans Liv og Værker*; and Aakjær (1903–04), *Steen Steensen Blichers Livstragedie i Breve og Akt-Stykker* (3 vols).
3 Whitney (1924), English primitivistic theories of epic origins, pp. 337–78.
4 Thomson (1951), *The Gaelic sources of Macpherson's "Ossian"*.
5 Macpherson (1803), *The poems of Ossian, the son of Fingal* (orig. 1765).
6 Tombo (1966), *Ossian in Germany: bibliography, general survey: Ossian's influence upon Klopstock and the bards.*
7 Blicher S. S. (1920a), *Steen Steensen Blichers Samlede Skrifter* (J. Aakjær & H. Ussing, eds), vols I & II: *Ossians Digte.*
8 Blicher N. (1978). *Topographie af Vium Præstekald*, pp. 45–6 (orig. 1795).
9 Macpherson (1803), Preface: "A dissertation on the poems of Ossian," pp. 2–3, 17.
10 Blair (1803). A critical dissertation on the poems of Ossian the son of Fingal: in Macpherson (1803), pp. 127–8. On Blicher, Virgil and Ossian, see Nørvig (1943), pp. 30–2.
11 Ibid., p. 139. Blair writes here: "The contrast which Ossian frequently makes between his present and his former state, diffuses over his whole poetry a solemn pathetic air, which cannot fail to make an impression on every heart."
12 Ibid., p. 72.
13 Pontoppidan (1763). *Den Danske Atlas*, vol. I, p. 99: Macpherson (1803), Preface, p. 17.
14 Blicher (1920d), Jyllandsrejse i Sex Døgn (orig. 1817): in *Skrifter* (J. Aakjær & H. Ussing, eds), vol. IV, p. 120.
15 Ibid., p. 214.
16 One of the first collections of this "national romantic" poetry was edited by Blicher and included poems by these authors (cf. Blicher (1823), *Bautastene*).
17 Ingemann (1815). Til Ossians Fordansker, Steen Steensen Blicher, i Anledning af Hans Første Udgivne Digte, pp. 338, 340.
18 Blicher (1920d), pp. 215–16.
19 Ibid., p. 218.
20 Ibid., pp. 223–34.
21 Blair (1803), p. 67.
22 Macpherson (1806). *Ossian*, vol. II, p. 15: quoted in Whitney (1924), p. 354. See also, Blair (1803), p. 73.
23 Blair (1783), *Lectures on Rhetoric and Belles Lettres*, vol. II, pp. 313–14: quoted in: Whitney (1924), p. 356.
24 Blair (1803), p. 69.
25 Goldsmith (1773). *The deserted village*, pp. 18–19.
26 Blicher (1930), Min Tidsalder (orig. 1842): in *Skrifter* (J. Aakjær & J. Nørvig, eds), vol. XXVI, pp. 260–1.
27 On Scott and Scotland, see: Lukács (1962), *The historical novel* (H. Mitchell & S. Mitchell, trans.); McLaren (1970), *Sir Walter Scott the man and patriot*; and Muir (1936), *Scott and Scotland.*
28 Scott (1863), *The Highland clans* (orig. 1816), pp. 94–6.
29 Olwig (1974): also Olwig (1981a), Var Blicher den Vilde Hedes Eller Menneskets Digter?: in *Museerne i Viborg Amt*, vol. XI, (Marianne Bro-Jørgensen, ed.). pp. 88–95.
30 The nomad – between hunter and shepherd, like the forest Indian, or the mountaineer, was regarded as a stereotype of natural man at that time. On the symbolism of the desert and sea, see Auden (1967), *The enchafèd flood, or the romantic iconography of the sea.*

31 Blicher (1945a), The hosier and his daughter (orig. 1829): in *Twelve stories* (H. A. Larsen, trans.), Princeton: Princeton University Press, p. 220. Minor emendations in the text have been undertaken by K. O.
32 Ibid.
33 On Blicher's view of such economic speculators, see Olwig (1974). Blicher did not oppose afforestation when it did not interfere with agriculture – see Aakjær (1903), vol. I, pp. 259–303.
34 Blicher (1945a), pp. 220–21.
35 Olwig (1974), pp. 100–05.
36 Blicher (1924b), Digterens Lyksalighed (orig. 1829): in *Skrifter* (J. Aakjær & G. Christensen, eds), vol. XIV, p. 122.
37 Blicher (1945b), The robber's den (orig. 1827): in *Twelve Stories*, pp. 118–19.
38 Macpherson (1773), *Introduction to the history of Great Britain and Ireland*, 3rd. edn, p. 3.
39 Burnet (1965), *The sacred theory of the Earth*, pp. 109–10.
40 Nicolson (1963), *Mountain gloom and mountain glory*.
41 Blair (1803, p. 50) writes: "In the infancy of societies, men live scattered and dispersed in the midst of solitary rural scenes, where the beauties of nature are their chief entertainment. . . . Their passions have nothing to restrain them: their imagination has nothing to check it. They display themselves to one another without disguise; and converse and act in the uncovered simplicity of nature. As their feelings are strong, so their language, of itself, assumes a poetical turn."
42 Willey (1962), *The eighteenth-century background*, pp. 23–4.
43 See Chapter 1, n. 22.
44 Willey (1962), p. 101.
45 See Chapter 1, n. 20.
46 Turnock (1970), *Patterns of Highland development*.
47 Graham (1956), *Colonists from Scotland: emigration to North America, 1707—1783*, p. 58.
48 Whitney (1924), pp. 345–7.
49 Marx (1973). *Gundrisse* (M. Nicolaus, trans.), p. 83.
50 On the emergence of landscape in art, and its ideological importance, see: Barrell (1972), *The idea of landscape and the sense of place 1730—1840: an approach to the poetry of John Clare*; Barrell (1980), *The dark side of the landscape: the rural poor in English painting 1730—1840*; Daniels (1982), *Humphrey Repton and the morality of landscape*, pp. 124–44; and Williams (1973), *The country and the city*.
51 Tuan (1974), *Topophilia*, pp. 129–49.
52 Lovejoy (1973), *The great chain of being*, pp. 292–3.

4 *The politics of landscape*

The young Søren Kierkegaard, writing in 1838, remarked upon the political implications of Steen Steensen Blicher's conception of Jutland's nature. According to Kierkegaard, "a certain new beginning takes place" in Blicher's work, which is undertaken "on the strength of an entire positivity which, so to say, awakens and finds expression, and, with youthful freshness, renews and regenerates itself with autochthonous originality". Blicher, in Kierkegaard's eyes, gives expression to a totality which has "nature's profundity as its creative source". Kierkegaard's language challenges the reader, but his meaning is quite clear if it is placed in the context of a conception of nature as embryonic principle. Blicher finds his creative source not in nature as thing, but in nature as potential, ever renewing and regenerating itself. This potential however, is not a cosmic principle; it emerges "with autochthonous originality" from the Jutland soil. Kierkegaard concludes that

> here is a unity, which in its immediacy points meaningfully into the future, and which must necessarily capture much more of the present than it has done, and possibly thereby come to have a beneficial effect on the prosaic manner with which politics has been treated until now.[1]

Kierkegaard was right, Blicher did have a poetic impact upon politics, and a closer examination of his work reveals that this was quite intentional.

The first decades of the 19th century in Denmark were marked by military defeat, territorial loss and financial collapse. Denmark's wealth during the 18th century had derived from its ability to maintain political neutrality and thereby the neutrality of its large merchant shipping fleet. The Napoleonic wars, in which Denmark became allied with France, proved to be a disaster for the country and resulted both in the loss of the Danish fleet and Norway, and financial weakness during the years of depression which followed the armistice.[2] It was at this low point in Denmark's history that Blicher dreamed of national renaissance:

> I am something of an actor, and I have a role, even if it is only self-appointed, in a great historical drama: the rebirth of Denmark.[3]

This role required the poet to take the side of the downtrodden:[4]

> The true bard knows that he did not take the harp,
> In order, with its strings, to drown out the
> complaints,
> Which rise from the cottages up to the ruler's
> throne.[5]

Nevertheless, prudence dictated that the bard use his harp tactfully. The rural idyll patterned upon Virgil's *Eclogues* provided a useful "vaile" under the cover of which the poet could make his point.[6] As his biographer, Johannes Nørvig, has noted, Blicher found a useful literary model in Virgil who also "lived under an absolute monarch who did not allow the written expression of deviant political beliefs; for his poetry he had to choose neutral subjects such as his Bucholia, country idylls, which praise the era of peace created by Augustus, and Georgica, a panegyric to human labor in the service of society, with evocative pictures of Italian landscape".[7]

Blicher's use of the idyll can be seen in "The eve of the three Holy Days"[8] where he contrasts the character of the heroic peasant, Sejer (victory), who lives near the site of an ancient parliament or *ting*, to that of a morally corrupt feudal aristocracy. Superficially an adventure story in which Sejer captures the fearsome outlaws who have been terrorizing the community, it also supports Blicher's argument that the people are their own best guardian, a belief which sustained his campaign to arm the general populace for the purpose of civil defense. The story also articulates his more general concern with the democratization of Danish society, based on an enlightened and independent agrarian populace opposed to domination by either the feudal aristocracy or the increasingly powerful urban commercial interests.[9] Blicher hoped that the agrarian reforms recently promulgated by the monarchy would lead to a state in which "the poet's fabled golden age would become reality, and the saturnalian period would return in all seriousness to Denmark".[10] There was indeed something saturnalian about these turn-of-the-century reforms, which gave ownership, or life tenure, of the land to many farmers, removing them from feudal obligations and providing universal access to a basic education. Blicher visualized this world in the conclusion to "The robber's den," of 1828. All classes meet in a traditional spring festival held on a village common bordering the heath:

> Towards the west end of Aunsbjerg woods there is an open place, quite a good sized green surrounded by venerable beech trees. Every year, on the afternoon of Whitsunday, most of the people who live in the surrounding parishes gather there. Many houses are standing empty that day, or they are guarded only by the blind and the bedridden; for the lame and the crippled – provided they have their eyesight – must at least once a year enjoy the forest newly in leaf and

> bring home a light green beech bough – like Noah's dove – to the dark dwelling which is often a Noah's ark in miniature.
>
> What fun! What crowds! The horse pasture – for that is the name by which this gathering-place is known – is like an enormous beehive: constant stir, everlasting thronging back and forth, in and out, all busy only with sucking up the honey of joy, drinking in the exhilarating summer air. . . . Far from the great hive one hears an incessant humming and buzzing – the bees are swarming. As we draw closer, the noise grows louder and the monotonous mass of sound dissolves itself into cries, singing, laughter, snapping of beech leaves, music of fiddles and flutes. The crowd surges in and out of the green edges of the forest, the peasants in their Sunday best, the gentle folk in smart summer clothes, the gentlemen in black, the ladies in white . . .
>
> You are in the horse pasture. This is the Whitsunday vespers in Lysgaard district, the day of homage to beautiful and ever-young Nature, the leaves of the forest, the triumph of summer . . . [It] is surely an old custom, perhaps as old as the forest itself.[11]

The conclusion of "The robber's den" is a perfect example of the classical conception of nature adapted to a modern setting. Blicher depicts the physical character of spring, but it is pregnant with symbolisms of the birth and rebirth of society. We are reminded of Virgil's lines:

> Spring showers her leafy blessings on the trees,
> Spring clothes the woods; in spring the
> swelling earth
> Demands the seed of life.
> . . .
> The pathless brakes resound
> With birds in full song; cattle seek their kind
> On certain days; the gravid earth brings forth,
> And to warm Zephyr fields unbind their breasts;
> . . .
> Thus dawned I trow, the birthday of the
> world,
> And kept its even way. That time was spring.[12]

It is spring, but it is also the birthday of a new world, and Blicher reinforces the point with a series of symbols of social rebirth, from Whitsunday, to Noah's dove, to an ancient native fertility ritual. For Blicher the rebirth of nature symbolized the rebirth of society's own nature.

Blicher was more than a poet, he was a social activist. He modelled himself upon the ancient bard whose song aimed to influence the ideas of his society, but he also actively participated in the affairs of that society. The emphasis in the young Blicher's use of the Ossian esthetic was upon the restoration of the wild physical nature of the ancient past. The mature Blicher, by contrast, emphasized the restoration of a more civilized and democratic society which he believed to have existed in ancient times. This was given material expression in his attempt to realize the festival depicted at the conclusion of "The robber's den," in his initiation of the so-called Himmelbjerg (Heaven Mountain) festivals of the 1840s.[13] These festivals were intended to bring together Danes from all classes and all parts of the country at the summit of a heather-covered Jutland hill, which was popularly believed to be the highest in the country. The program consisted of an Arcadian mixture of song, poetry and patriotic speeches intended to enable the participants to "rediscover themselves under their varied conditions of life and especially to rediscover their fresh, pure and youthful spirit".[14] It was this personal rebirth that Blicher believed would lead to the rebirth of the beloved fatherland. It is a measure of the success of these festivals that a contemporary literary critic pronounced them to be Blicher's "very best poetry," and that they became the model for similar festivals which have played a social and political role in Denmark up to modern times.[15]

The festivals may have been Blicher's best poetry but they eventually aroused the displeasure of the authorities. Popular assemblages stressing such heady themes clearly were not designed to make the authorities sleep easily, and the content of the speeches confirmed their fears. An oration by the social critic, Carl Plough, quickly led the district governor, Hans Lundholm, to call for prior censorship. A call which he justified in his report on the festival:

> After the speaker had made general remarks to the effect that things were not as they should be in the fatherland, he proceeded thence to a comparison of the past and present in which for the most part he spoke as someone who is offended by lost values, and who presents the present as deplorable when compared to the past. It was then he claimed that the peasant stood in parliament beside the king, when the will of the people also was the king's will. Now, by contrast, the farmer's or the burgher's wishes did not always even reach the ear of the king . . . – By the way, he also gave his speech the turn that it was not the government he was trying to blame or make the cause of the country's shortcomings, but rather the people themselves, who had the power to obtain a strong and free government. He also mentioned the often-discussed theme of the bureaucracy as a barrier between the king and the people. Finally, he encouraged the people to struggle with force and independence for a better future for the fatherland. He concluded with a greeting to that future which was to be accompanied by a "Hurrah" which was capable of wakening the sleeping and frightening the cowardly.

Figure 4.1 A view from the Himmelbjerg on a foggy day.

> I am convinced that this speech, far from missing its purpose, has left a deep impression upon the crowd, and I therefore see it as my duty to make his excellency aware of its content.[16]

The festivals promoted the image of an innocent seasonal gathering, but they succeeded in fulfilling George Puttenham's claim for the pastoral: "under the vaile of homely persons and in rude speeches to insinuate and glaunce at greater matters".[17]

Blicher himself realized that the national rebirth of which he dreamed would not come about without the introduction of more democratic forms of government. Thus, while more discreet than Plough, he also played upon the theme of the regeneration of a democratic golden past. He opened the first festival, held in August 1839, with a speech in which he spoke at length about olden times when the people met in a similar fashion to hold parliament at the *ting*. He contrasted this with a description of the "dark centuries" after such meetings had ceased. Then he came to his theme of a modern national "awakening" of the people, symbolized by the rebirth of nature in the spring. Evidence of its renaissance was the first meeting of the proto-democratic "Assembly of the Estates" on May 28, 1831. Even though it only had an advisory capacity, the assembly signalled an important change in Danish governmental policy:

> We will celebrate the 28th in the month of flowers as the rebirthday of

the fatherland; why shouldn't this political resurrection be celebrated on this mountain, which bears the name of heaven?

I see before me these dark banks, decorated with the creator's most beautiful flowers – Danish men and women, greeting the May sun as it rises in the east. I hear them greet it with songs – with freedom's folksongs. The Danish beech, the Danish waves echoe the jubilant tones.[18]

Blicher and the perception of Jutland

The bard was warrior as much as poet, active patriot as much as reflecting spectator. Blicher himself was more than poet; he sought actively to improve social life. Although with little success, he tried his hand at heath cultivation and afforestation; he wrote on agronomic and geographical topics, and he maintained active membership in agricultural improvement organizations. His 1838 geography of Viborg County remains an important historical source of conditions in the region.[19] But his main impact was through his poetry and fiction; he was a visionary rather than a practical man. There is no doubt that the national attention which he attracted through the Himmelbjerg festivals and his writing did much to boost the local Jutland sense of identity. He not only made the Himmelbjerg a national symbol, he put Jutland on the map.

It may seem odd that in a country the size of Denmark there would remain areas as late as the 19th century largely unknown to all but local residents. Size alone, however, is not the determining factor here. Denmark is an archipelago which even to this day is linked in large measure by sea routes. Because of its excellent harbors the eastern coast of Jutland has always been relatively accessible. Even so, given the vagaries of the weather, a trip from Copenhagen might take several days. The western coast, with poor port facilities, shallow waters and few natural harbors, was not very accessible by boat, particularly as this required a dangerous journey around the northern tip of Jutland. Overland the journey was hindered by poor roads over sandy heaths and marshy bogs. The difficulty of travel is indicated by the fact that Blicher, who enjoyed travel, was 51 years old when he first saw the North Sea, though his home in mid-Jutland was less than 100 km distant from the coast.

Rural Jutland was not only physically distant from Copenhagen, it was also mentally distant. Denmark had been a model of enlightened despotism since 1660, strong enough to survive almost two centuries. It was a highly centralized, hierarchical, bureaucratic state, which patronized urban mercantile interests. Geographically, it was the capital, Copenhagen, which benefited most from absolutism. It grew into a primate city, a city swollen out of proportion to other urban areas. Robert Molesworth was right when he stated that the monarchy favored foreigners over natives. It invited financially astute Germans and gave them newly created aristocratic titles; it brought in technical and economic expertise to develop manufacturing, and artists, artisans and hardworking immigrant groups such as Jews and Huguenots. It created a cosmopolitan mercantile society in which the traditionally rural-dwelling Danes, and their language, took second place. The scheme for German colonization of the heaths based on large, strictly supervised nucleated settlements was a typical product of the attitude of Danish enlightened despotism. People like Hans de Hoffman and books like *Den Danske Atlas* represented an emerging national reaction gathering strength in the late 18th century and reaching maturity in the early 19th. This was manifested in the recognizably national romantic art of B. S. Ingemann, Adam Oehlenschläger and their Jutland counterpart, S. S. Blicher; and in the peaceful revolution of 1849 which resulted in constitutional democracy and the economic and cultural renaissance of the provinces. Copenhagen began to take an interest in the countryside, and distant Jutland, the repository of the ancient Danish language and heritage, began to arouse people's curiosity.

Through his novels and stories Blicher created a national interest in the Jutland heaths, and others followed with publications on social and agronomic conditions in the area. Among these was Pastor Frederik Carl Carstens, whose stated admiration of Blicher's writing is readily apparent in the younger man's landscape descriptions. Lamenting the destruction of the forests he believed had once covered Jutland, Carstens adopts Blicher's elegiac style to describe the heath: "dressed in dark, quite as if it were mourning our forefather's deeds, when they so willingly helped steal the beauty it had inherited for eons".[20] Carstens was less a poet than a poetaster. For him the tree was not simply a symbol of past fertility, it was the very means by which the "water veins" beneath the heath would be made to burst forth once again.[21] He believed that local landscape taste had to be changed if the local population was to reclaim the heathlands for cultivation. Carstens was a native of the fertile south-east Jutland island of Als, an area noted for its hedge-enclosed fields and farms. He became involved in the social and economic problems of heath cultivation through his assignment, in the period 1832–45, to the pastorate at Frederiks on the Alheath, one of the few surviving areas of concentrated German settlement. His own background, and his observation of the radically different landscape tastes of colonists and native inhabitants, made him acutely aware of the role of landscape taste in landscape change. He was struck by the native population's lack of interest in hedges and trees as shelter belts for their farmsteads. It seemed to him that they aimed to remove any vegetation which would break the open sweep of the windswept plain. In the interest of agricultural improvement he promoted the beauties of enclosed landscape of the type which supposedly characterized the golden age prior to

Figure 4.2 The frontispiece from Blicher's early collection of national romantic poetry, *Bautastene*, or "monoliths", which memorialized Danish heroes. According to Hugh Blair, (1803, p. 66) "The great object pursued by heroic spirits, was 'to receive their fame,' that is, to become worthy of being celebrated in the songs of the bards: and 'to have their name on the four grey stones' ".

deforestation. Carstens' was a more pragmatic vision than Blicher's. Rather than the return of a "saturnalian period," with diffuse notions of social reform and democratic freedom, Carstens looked to the improvement of the Jutlander's diet that sheltered vegetable gardens would allow.[22] Though he criticized their landscape taste, his loyalty was nevertheless to the native population. He was all too aware of the failures of the colonization scheme, reporting a dramatic population increase in the colony out of proportion to its monocrop potato resource base, and he favored the native mixed agricultural use of local resources.[23]

Blicher had taught Carstens to see the heath as landscape, and to appreciate the importance of the relationship between landscape taste and landscape use. Carstens also took Blicher's theme of substituting the peaceful heroism of the farmer for the heroism in war of the soldier. In his view, cultivation of the heaths would be adequate compensation for the military loss of territory:

> Norway was ripped out of our hands, and it was thought that Denmark's strength was thereby paralyzed. This, however, does not mean very much as long as our country hides in its interior many extensive heaths which, if properly used, would be able to raise and nourish just as many strong arms in Denmark's heart as are encompassed by that country in the North Sea.[24]

Carstens was not the first to make this suggestion. As early as 1800 the Danish poet and economist, Christian Olufsen, had urged the Danes to follow the example of those in Britain who were calling for internal colonization of wastelands as a substitute for a financially wasteful external imperialism.[25] Olufsen argued that the heath should not be colonized from without, but cultivated and afforested by its native population, thereby restoring its former state. Olufsen, in turn, had an early influence on Blicher, it was he, in fact, who introduced Blicher to the poetry of Ossian. The notion of restoring former national glory by turning attention from external possessions to the reclamation and intensification of the nation's internal wastelands occurred elsewhere in Nordic literature at the time. The Swedish poet Esaias Tegner's gothicist classic *Svea*, published in 1811, two years after the Swedish loss of Finland to Russia, called upon the nation to "conquer Finland again – within Sweden's boundaries".[26] Unfortunately for the Danes the Swedes did not abandon external territorial expansion, and three years later Sweden annexed Norway from Denmark. Now it was not so much Sweden as Denmark which needed to recoup external loss within the national borders. Carstens was one of the first to argue that this should be done through the cultivation of the heaths.

Carstens's writings struck a sympathetic chord in a fellow native of Als, Colonel Hans Christian Riegels, a man with talents both as a practical forester and as an organizer and agitator.[27] Stressing the patriotic importance of afforestation, he initiated forestry programs on the heath which enjoyed both state and local support, particularly in the Viborg area. An ancient capital, now a provincial town, Viborg was surrounded by heaths, but it provided fertile ground for those who proselytized for the local and national patriotic importance of heath reclamation and afforestation. Viborg had much to gain from the intensification of local agriculture. Its burghers saw the possibility thereby of regaining something of its former importance as a regional centre. In short, Viborg was open to the patriotic arguments of Riegels and Carstens, and by the mid-19th century it had become a center of interest in heath reclamation. Here a whole circle of civic leaders invested time, effort and capital in afforestation and cultivation projects, receiving local support from a newly emergent class of independent farmers.[28] Significantly, these were often the same people who had supported Blicher's Himmelbjerg festivals. Land improvement went hand-in-hand with more grandiose ideas for the restoration of former social conditions in which the common man played an enhanced role in the social and economic life of the country.

Among the earliest, and most successful, of those who urged the planting of trees on the heath for patriotic reasons was the clergyman Hans Bjerregaard. A neighbor of Blicher and Carstens, he was the son of one of 18th-century Denmark's most prominent peasants, Hans Jensen Bjerregaard. Hans Jensen had risen from near serfdom and illiteracy to farm ownership and leadership within his estate. Pastor Bjerregaard naturally believed in the potential of the farming class. At his own expense he provided the heath farmers, including Blicher, with free saplings in order to encourage the planting of sheltering trees where the land was not suitable for cultivation. Like Blicher, Bjerregaard believed that the world was approaching a new golden age in which land reclamation would play an important part.[29] Though it is true that he disagreed with Blicher over practical matters of tree planting, he appreciated the significance of Blicher's national romantic poetry.[30] His taste for romantic landscape was expressed in his romantic garden, decorated with monoliths recalling ancient national glory.[31] Bjerregaard knew how to put his romantic notions into practice and he had a remarkable success convincing the local populace to plant trees.

The circle of people which developed in Viborg and its environs in the first half of the 19th century met, by and large, with only limited success in their efforts to reclaim the heaths. They needed greater expertise, experience and, above all, capital, but they succeeded in developing the local basis of support for improvement.

The national perception of the heaths

In 1850 the heathlands were thinly populated and economically poor. It was difficult even for an active local populace to make major changes in the economy of such a peripheral area, even had economic and demographic conditions been favorable. Agricultural intensification required a new transport infrastructure. The traditional grazing-based economy yielded animal products which either walked to market or, like wool, could be transported cheaply. The road network was notoriously poor and in many areas only a native could navigate the vast stretches of sandy ruts which

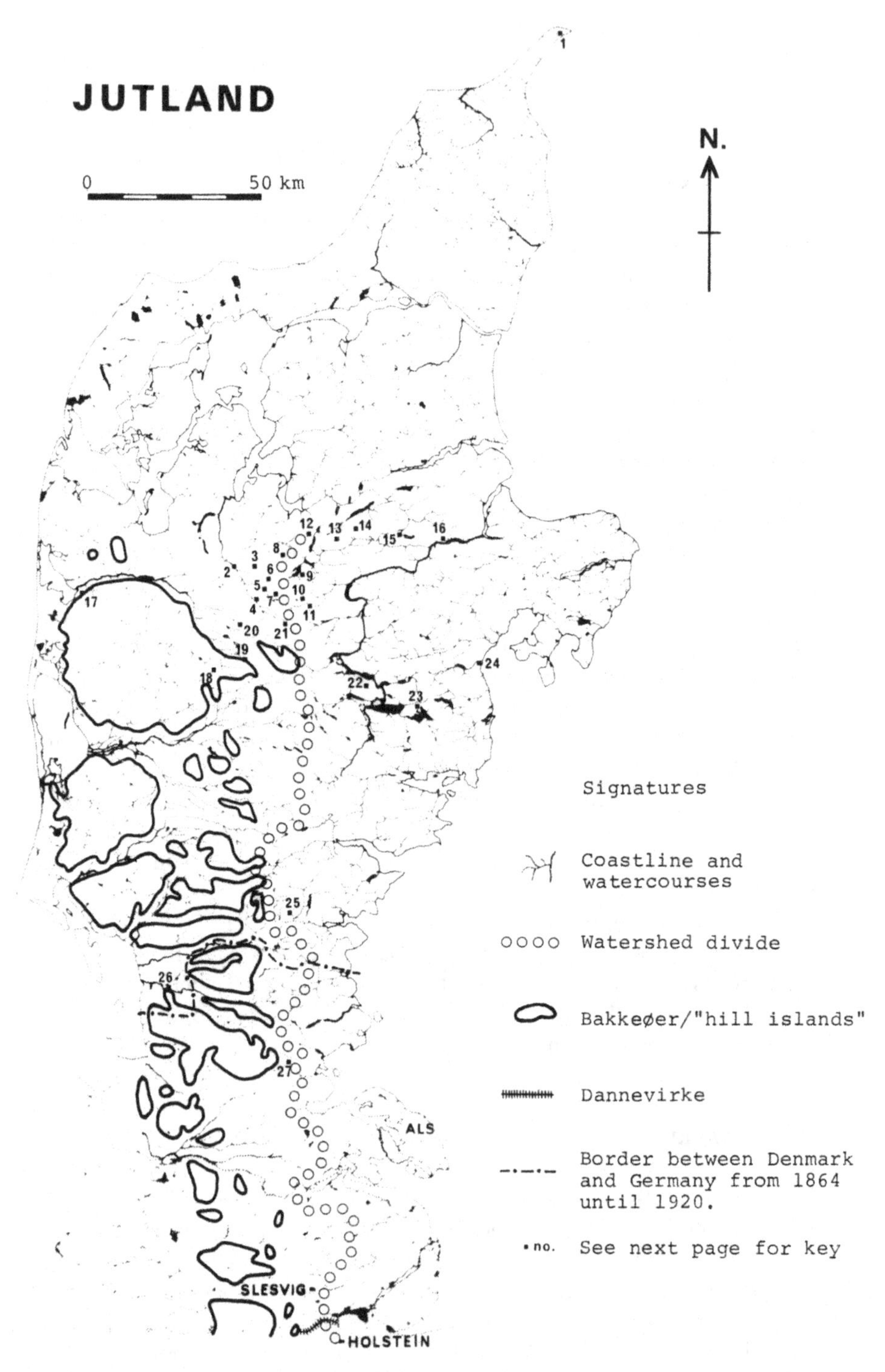

Figure 4.3 A map of Jutland showing location of place names mentioned in the next two chapters (map by Niels Hansen).

criss-crossed a featureless terrain. In order for investment in more intensive forms of agriculture to be profitable the farmer had to be able to transport goods, like grain or potatoes, in bulk over the long distances to market. Infrastructural improvements of this magnitude in such poor and thinly populated districts required state support to redirect economic resources from more wealthy districts. Such support was now no longer a matter of royal decree; under the constitutional monarchy, it required legislative approval, and hence broad national support. The problem was exacerbated by the frequent financial paralysis of government, engendered by political and constitutional controversy over the remainder of the century. It was here that Blicher's literature and national prominence could play an important role in awakening interest in heath reclamation. Blicher's influence at a national level can be illustrated through the example of two of 19th-century Denmark's most prominent authors, Hans Christian Andersen and Meir Goldschmidt.

Hans Christian Andersen and Meir Goldschmidt

Andersen is best known today for his fairy tales, but in his day he had an international reputation as a novelist and essayist. As early as 1830, inspired by Blicher's stories, he visited Jutland. Bad weather cut the trip short, but he repeated it in 1859 at the invitation of A. E. M. Tang, the owner of the West Jutland manor of Nørre-Vosborg. Tang actively supported both heath reclamation and greater independence for the region's peasantry. He was among those who had helped Blicher organize the Himmelbjerg festivals and was the poet's closest confidant in the years up to Blicher's death in 1848. Tang gave a huge party for Andersen, attended by the local peasantry. Of them Andersen wrote:

> They are a magnificent race, farmers with refinement and the desire to know and understand.

1 Skagen
2 Karup Creek
3 Kongenshus Memorial Park
4 Karup village
5 Al Heath
6 Grønhøj village
7 Havredal village
8 Finderup village
9 Hald Lake
10 Vium village
11 Aunsbjerg Manor
12 Viborg
13 Tap Heath
14 Ørum village
15 Foussingø village
16 Randers
17 Nørre-Vosborg Manor
18 Herning
19 Ikast
20 Ilskov village
21 Bækgaardsdiget (Grathe Heath)
22 Himmelbjerget
23 Skanderborg
24 Aarhus
25 Leirskov village
26 Ribe
27 Houslund village

He also was impressed by Tang's initiatives:

> the countryside itself will be a grain and forest land, I am convinced of that; but then, the romantic, heather-covered heath will be gone, with all its loneliness, its mirages, and its character of an ancient time.[32]

The impressions from this trip formed the basis for one of Andersen's most popular poems, "Jutland between two seas" (1859). This poem contains all the characteristic elements of Blicher's gothicist perception of Jutland: the graves of ancient heroes, the rune stones, the sublime landscape, the approaching golden age:

> Jutland between two seas
> like a runic cipher laid,
> The runes are giant graves
> within the splendor of the forest
> and the great solemnity of the heath,
> here lives the desert's mirage.
>
> Jutland you are the chief land,
> a highland with forest loneliness:
> Wild in the west, with sand duned cliffs,
> rising in place of mountains.
> The Baltic and the North Sea's waters
> Embrace across Skagen's sands.
>
> The heath, yes, it is hard to believe –
> but come yourself, look it over:
> the heather is a splendid carpet,
> flowers crowd for miles around.
> Hurry, come! in a few years
> the heath a grainfield will be.
>
> Between wealthy peasant farms
> soon steam dragons will fly;
> Where Loke now drives his herd,
> forest will overgrow the land.
> The Briton will fly across the sea,
> and visit Prince Hamlet's grave.
>
> Jutland between two seas
> as a runestone it is laid.
> the past is spoken by your graves,
> the future unfolds your power;
> the sea with all its breath
> sings loudest of Jutland's shores.[33]

In 1866 another of Denmark's leading literati made what he called a "pilgrimage to Blicher's poetic land" and chronicled his visit in a journal published under the title, *A heath journey in the Viborg region.*[34] Meir Goldschmidt, a Copenhagen Jew, made his international reputation with the novel, *A Jew*, and the satirical jibes at Søren Kierkegaard in his journal *Corsaren*. In Denmark he is remembered as one of the mid-century's most important literary figures and as a sharp social critic. His heath pilgrimage came at a turning point both in Goldschmidt's life and in the life of the nation. In 1864 Denmark was defeated by Bismarck's Germany and lost the south Jutland provinces of Slesvig-Holsten. This loss cut deep not only because of the value of the territory, but also because a considerable Danish population lived in Slesvig. Only in 1920 was the northern portion of Slesvig reunited with Denmark. In the years prior to the war Goldschmidt had been active as a journalist, and in this capacity had agitated unsuccessfully for an aggressive military policy. In frustration, he abandoned Denmark for England where sales of *A Jew* had gained for him a reputation as a novelist. He returned to Denmark in 1863 and in this time of crisis made his 1866 journey to the heaths. In 1867 he took up a new career as a dramatist and prosaist gaining recognition as a serious author. Goldschmidt was a complex figure, an ardent patriot, but also a Jew who felt deeply his people's alienation in a Christian country. The trip to Jutland helped him regain his sense of national identity.[35] His journal is worth examining in this context because it helps explain how, at this time, the reclamation of the heaths came to provide, for many, a means by which the nation itself could regain the collective identity which had been shattered by humiliating defeat.

Goldschmidt's journey was very much a Blicheresque pilgrimage. He wrote:

> I love the aforementioned districts with a sensibility in which is blended something which lies beyond the senses. It is not just the countryside and the people, which, insofar as I know it in all its simplicity, is singular, varied and to me attractive, but it is also the dead, and that which never has been which attracts . . . With the word "heath," especially referred to the heaths between Viborg and Karup Creek, something invisible rushes by me – by most of us I could well say – which we reach out after: the Blicher characters.[36]

Goldschmidt sought out barrow-topped vantage points where he could find "an uncommon view" of "the regions where *Blicher* composed and localized his most poetic figures".[37] It was not, however, just the literary landscape which he sought out, but the people and places from which Blicher created that landscape:

> I spent some time in aimless joy over the fact that now I really was in the country, in the middle of Jutland, alone with folk who are outside the bourgeoisie, outside intellectual interests, interests which for others like myself are a necessity and a strain; alone between those naive interests which seem to us to lie closer to life's truths and freshness, and which give a peaceful heart and long life.[38]

The heath and the nation

Blicher's landscape to which Goldschmidt made his pilgrimage bore the lineaments of a Virgilian nature embodied in the rural idyll. Goldschmidt was a sophisticated urbanite, tired of the city's unnatural ways, who journeyed to a distant peripheral place in order to experience a more natural form of society. Here he found a people who preserved the customs of a distant, more natural past, and who could recount this past in poetic tales. In his journey he rediscovered his own inner nature and that of his country. Physical nature – the landscape itself – is clearly secondary to its symbolic associations with the proper way in which people should live.

In a sense, Blicher's landscape constitutes a negative pastoral. It is the ruin of the golden age landscape, incorporating the embryo of a regeneration of the ideal qualities that characterize the youthful state of the nation. Goldschmidt shared this perception of the heath:

> Like a destructive stormwind, time has swept over this place, and has carried away nearly everything which has emanated from the hand of man, taking with it the forests which once stood in this region. With this recollection in mind, the wavy hillocks give the impression of a terrain which has been chastized and suppressed. Only by the water mill, a little down stream from the Karup church, can a little patch be found which seems to have been spared: a very little thicket on the green meadow floor.

In this place Goldschmidt describes his meeting with a small group of men who have been staying in Karup, including a geologist, Professor Johnstrup, and Enrico Dalgas, who is identified as the head of a newly formed society dedicated to the reclamation of the heaths. They had met here, we are informed, in order to investigate the possibility of achieving their goal of renewal by damming streams and creating artificial meadows. This would expand the supply of winter fodder and, thereby, the fertilizer available to farmers sustaining the traditional form of heath-land agriculture. The scientist, Johnstrup, we now know, was to later revolutionize Danish perception of the physical geography of the heaths, and Dalgas, who led the Heath Society, from its formation in 1866, was to lead a successful national drive to reclaim the heaths. Goldschmidt like Andersen and Blicher before him, had his finger on the psychological pulse of the nation and recorded the following prophetic thoughts:

> The district is, as I have said, desolate and suppressed; but our time has a liberating power; it raises that which has been broken; mediates and reconciles even the forces of nature, bringing them into partnership with humanity. I had the luck, at just this spot, to meet representatives of that tendency and that ability in our time.[39]

Notes

1 Kierkegaard (1906), *Af en Endnu Levendes Papirer*, pp. 60–1.
2 For historical background, English language readers are referred to Oakley (1972), *A short history of Denmark*.
3 Blicher (1933), Letter to Peter Christian Koch (orig. 1845): in *Samlede Skrifter* (J. Nørvig, ed.), vol. XXXII, pp. 163–4.
4 Olwig (1974), Place, society and the individual in the authorship of St. St. Blicher, pp. 69–114.
5 Blicher (1920d), *Jyllandsrejse i Sex Døgn* (orig. 1817): in *Skrifter* (J. Aakjær & H. Ussing, eds), vol. IV, p. 172.
6 Olwig (1981b), Literature and "reality": the transformation of the Jutland Heath, pp. 47–65.
7 Nørvig (1943), *Steen Steensen Blicher, Hans Liv og Værker*, pp. 32–3.
8 Blicher (1930a), De Tre Helligaftener (orig. 1841): in *Skrifter* (J. Aakjær & J. Nørvig, eds), vol. XXVI, pp. 1–15.
9 Aakjær (1903–4), *Steen Steensen Blichers Livs-Tragedie i Breve og Akt-Stykker*, vol. II, pp. 262–91.
10 Blicher (1924a), Danmarks Nuværende Tilstand (orig. 1828): in *Skrifter* (J. Aakjær & G. Christensen, eds), vol. XIII, p. 179.
11 Blicher (1945b), The robber's den (H. A. Larsen, trans.), pp. 118–19.
12 Virgil, *Georgics* II: 386–402: in Virgil (1946), *Eclogues and Georgics* (T. F. Royds, trans.), pp. 106–7.
13 The festivals were held annually from 1840 to 1844. They inspired similar festivals elsewhere, particularly those at Skamlingsbanken in south Jutland in 1843–4, which were of importance for the development of a Danish national identity in the area. On the festivals, see Aakjær (1903–4), vol. II, pp. 292–424.
14 Blicher (1931). Bordtale ved Himmelbjergfesten 1843: in *Skrifter* (J. Nørvig, ed.), vol. XXVII, p. 240.
15 Barfod (1853), *Fortællinger af Fædrelandets Historie*.
16 The letter is quoted in full in: Aakjær (1903–4), vol. II, pp. 396–7.
17 See Chapter 1, n. 14.
18 The speech is quoted in full in: Aakjær (1903–4), vol. II, pp. 302–4.
19 Blicher (1928), Viborg Amt: in *Skrifter* (J. Aakjær & J. Nørvig, eds), vol. XXI, pp. 199–268, and vol. XXII, pp. 1–240.
20 Carstens (1844), *Bemærkninger over Heden og dens Træplantning*, p. 26.
21 Ibid., p. 35.
22 Ibid.
23 Carstens (1839), *Bemærkninger over Alheden og dens Colonier*.
24 Carstens (1844), p. 74.
25 Olufsen *et al.* (1800), Efterretninger om Alheden og Randbølheden i Nørre Jylland, p. 99.

See also Olufsen (1811), *Danmarks Brændselsvæsen, Physikalskt, Cameralistisk og Oekonomisk Betragtet*, pp. 301–38.

26 Tegner (1817), Svea: In *Svenska Akademiens Handlingar ifrån År 1796*, no. 6., p. 162.

27 For Riegel's views, see: Riegels (1847), Brudstykker fra en Udflugt over Hannover, Brunsvig og Magdeborg, især med Hensyn til Træplantning og Havecultur, pp. 76–128; and Riegels (1848), *Til Træplantningens Fremme i Almindelighed dog fornemmelig paa Hede=og andre slette Jorder der kun egne sig til Skovcultur.*

28 On early heath reclamation attempts, and the Viborg role, see: Oppermann (1889), *Bidrag til det Danske Skovbrugs Historie 1786—1889*; and Skrubbeltrang (1966), *Det Indvundne Danmark*, pp. 13–104.

29 Bjerregaard (1840), *Er den Evige Fred en saadan chimære og Umulighed, som en Æret Forfatter Nylig har Yttret?*

30 On Blicher and Bjerregaard, see Aakjær (1903–4), vol. I, pp. 259–303.

31 On Bjerregaard's garden, see Molbech (1824). Optegnelser paa en Udflugt i Jylland i Sommeren 1828, p. 154.

32 Andersen (1951), *Mit Livs Eventyr* (H. Topsoe-Jensen, ed.), vol. II, p. 106: also ibid., p. 212.

33 Andersen (1964), Jylland Mellem Tvende Have, pp. 714–15. According to Danish legend Hamlet was a Jutlander. Shakespeare, of course, identified Hamlet with the castle at Elsinore on Zealand. The poem was written while Andersen was *en route* between Randers and Viborg on a visit to the then 93-year-old Hans Bjerregaard (see Andersen 1951, p. 451).

34 Goldschmidt (1954), *En Hedereise i Viborg-Egnen*, (V. Andersen, ed.), p. 60.

35 On Goldschmidt, see Rhode (1967). Idedigtning og Politisk Gennembrud, pp. 95–701.

36 Goldschmidt (1954), pp. 10–11.

37 Ibid., p. 37.

38 Ibid., p. 47.

39 Ibid., p. 87.

5 *The "heath cause"*

The Danish defeat in 1864 did not resolve the Slesvig-Holsten issue. It might have done so had Slesvig been divided along cultural lines, but the annexation left a considerable Danish population in northern Slesvig, which was now within Germany. The defeat, and the uncertain international situation, led to calls by political conservatives for increased defense spending. The political left opposed this, and pressed rather for social and political reform. The right was dominated by the military, the major land owners, international merchants and the central bureaucracy – an inheritance from the period of absolute monarchy. Broadly speaking, the left was dominated by farmers, local merchants and the intelligentsia. The left controlled the lower house of the legislature while the right controlled the upper house and the executive. Conflicts between right and left were so severe that the different branches of government often failed to reach a compromise, and the ensuing governmental paralysis led to rule by executive fiat. Danish national identity was thus threatened not only by military defeat and the consequent loss of about one-third of its territory, but also by debilitating class conflict and political deadlock. Matters were further complicated by the fact that cheap imports of foreign grain into Europe put great pressure on the local agricultural economy. Heath reclamation, however, provided an issue around which all sides could rally.[1] It was Enrico Dalgas who was to bring this about by making reclamation a national cause.

Enrico Dalgas

Enrico Dalgas (1828–94) first became acquainted with the heaths in 1852 as a young army road engineer working in the Viborg area. This work gave him first-hand knowledge of the region's physical and cultural geography. It also brought him into contact with a circle of ardent supporters of heath reclamation, particularly Georg Morville, a young lawyer with an interest in natural science who was then engaged in an afforestation project with Hans Christian Riegels. These personal contacts, together with his personal experience on the heath, stimulated Dalgas's interest in reclamation. He observed how farms followed in the wake of the roads that he built, how farmers were able to dam streams and expand their meadows, and he saw that trees could be grown on the heath. Initially, however, other matters were to absorb all his attention. As an army engineer he had been involved in works to strengthen the "Dannevirke," an ancient line of fortifications across Slesvig, forming a bulwark against the south. He felt that too little was being done too late by way of defense and in 1863 he had the bitter task of helping lead the Danish retreat from this symbolic, but indefensible, bulwark in the face of Bismarck's advancing Prussian army.[2] A year after the end of war, in 1865, Dalgas published two articles in the leading Copenhagen newspaper. In the light of the recent national loss he called for a large-scale state project to cultivate and afforest the heaths.[3] The official reception to his articles was so cool that Dalgas decided to "go to the people of the heaths themselves".[4] Morville and a number of other interested persons helped organize a society to lobby for and coordinate a national project with the goal of "making the heaths fertile".[5] On March 28, 1866 The Danish Heath Society was formed under the leadership of Dalgas, Morville and Ferdinand Mourier-Petersen, a wealthy manor-owner with good connections in the ruling Conservative party. The society had a broad-based membership drawn from across the country. The Society's central objective, and the idea which attracted national attention, was later summarized in the motto:

> For every loss, compensation can be
> found again,
> What is lost without, must be
> won within.[6]

It was some time before the program so neatly summed up in these lines was formulated and put into practice. They were to become a symbol of the principles behind the development of modern Denmark, but initially the Heath Society had to contend with powerful opponents.

A major problem was the common belief that the heaths could be neither cultivated nor afforested. As Dalgas later wrote:

> There was very little belief in 1866, either in the Executive, the Legislature, or amongst the people, that it was possible to create forests on the heaths; and there was little confidence that the Heath Society would be able to accomplish anything of this sort. [But] . . . little by little, through lectures, publications and by beginning to plant here and there, it proved possible to create a favorable climate for this important cause.[7]

From an early date the Society recognized the need to influence public opinion. Its original program was summarized in seven points:

(1) To awaken interest in, and further education about, the heath and its treatment;
(2) to produce a practical geography of the heath, especially with regard to

its size and fertility in different regions;
(3) to study the possibilities for utilizing heathland water courses for irrigation;
(4) to remove hinderances to enclosure, etc.;
(5) to hasten the improvement of transport infrastructure;
(6) to promote tree planting on the heath by the establishment of tree nurseries, and the free distribution of saplings;
(7) to gain the support of the government and to work together with the agricultural societies.[8]

The last point on the Heath Society's program was especially important because one of the major opponents of its project was the prestigious Royal Agricultural Society (*Kongelige Landhuusholdningsselskab*), particularly its young leading light, Lieutenant J. C. la Cour. Lt. la Cour felt that Denmark's limited capital resources should be invested in the intensification of agriculture on the better soils of eastern Denmark. The heaths, he argued, were simply too isolated and too poor in soil quality to make the Heath Society's project profitable. A limited portion of the heaths, he contended, should be reserved for afforestation, but otherwise they were not worth the necessary investment.[9] The idea of some afforestation on the heaths was popular with the Conservative government and gained considerable support amongst wealthy manor owners, many of whom felt economically hamstrung by restrictive Danish forest legislation. This legislation made it illegal to reduce forest acreage, and much manor acreage was traditionally tied up in forest. As a means of circumventing this legislation they championed the idea of transferring forest land to Jutland, thereby freeing their own soil for cultivation while leaving constant the nation's total percentage of forest. This idea reflected a general pattern in Europe where the rise of economic liberalism had led to an attack upon forest protection measures.[10]

The Heath Society not only faced opposition based on economics, it also had to contend with arguments taken from the natural sciences, which were also effectively used by la Cour. In his influential book *Denmark's geognostic conditions* of 1835 the renowned Danish geologist, Johan Georg Forchhammer (1794–1865) had claimed that "the western portion of the country . . . is a great, unbroken plain," and had labelled it "hardpan flats". His accompanying colored geological map gave visual support to this notion by depicting the heathlands as one undifferentiated hardpan plain.[11] Forchhammer had little concrete evidence upon which to base this sweeping generalization, but he did have a preconceived notion published in an article with the revealing title: "On the great flood which struck Denmark in very ancient times". According to Forchhammer, the Gothic invasions of Rome were a consequence of the same natural causes which created the hardpan flats:

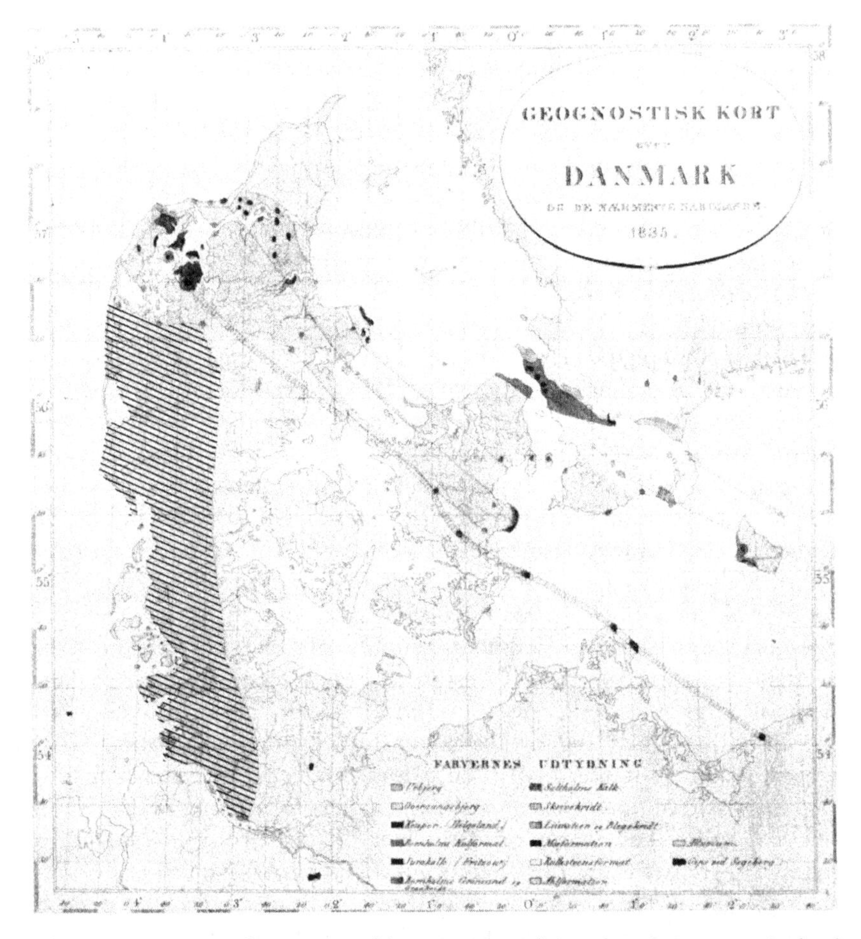

Figure 5.1 A map from *Danmark's geognostic conditions* showing western Jutland as an unbroken hardpan formation (diagonal hatching) (Forchhammer, 1835).

> Roman authors who discuss the Cimbrian and Teutonic invasions into Italy give as the reason for the immigration of these peoples the occurrence of a great flood which had disturbed the country. But already, long before, in the time of Alexander the Great, the Greek authors, Ephorus and Klitarchus, speak of a flood which was supposed to have disturbed the northern countries. This flood is known under the name, the *Cimbrian flood* . . . .[12]

According to this theory, not only had a flood caused the Gothic invasions; flooding also was responsible for the destruction of the Jutland environment because the layer of hard mineral matter which it had deposited prevented the growth of trees and paved the way for the development of the heath. Forchhammer's work was damaging to the Heath Society's program both because it gave the impression the heath was one vast infertile plain and because it suggested that the poverty of the region was environmentally determined.[13] It challenged the view that the heaths were essentially the outcome of social conditions and thus reparable through alteration of those conditions. The Heath Society knew that the heathlands' landscape was differentiated and not uniformly underlain by hardpan. They also knew that considerable progress had already been made both in agricultural intensification and afforestation. Theirs was not so much a practical difficulty, as one of public awareness and understanding. This is why the Heath Society placed such early emphasis upon lectures, geographies, their journal and other publications as a means of spreading information and altering public perception of the heath.

Dalgas, Schouw, and the Heath Society's program

The Heath Society was led by a triumvirate consisting of Morville, Mourier-Petersen and Dalgas. But in the public mind it was Dalgas who was identified with the cause of heath reclamation. Dalgas was the man seen in the field, conversing with farmers, and organizing projects. He was the skilled publicist whose lectures and writings shaped the Danish image of the heathlands. His practical success can be understood in the light of his background as an engineer on the heaths. The origin of his skills as a publicist is not so apparent, but would seem to stem from the influence of his uncle by marriage, Joachim Frederik Schouw, the internationally respected botanist and geographer.[14] When Dalgas was only seven, in 1835, his father died, leaving the family destitute and Schouw came to exert an important influence on the upbringing of his nephew. On the death of the father he took in Enrico for a time, and saw to it that he received a free education at one of Copenhagen's better private schools (as early as 1830, Schouw had sought to help the financially troubled family by taking Enrico's brother, Carlo, into his home). Enrico greatly admired his uncle, not least for his popular nature studies and geographic works, which he recognized as "capable of being understood by everyone," and which he felt "contributed much to the spread of knowledge".[15] Dalgas later attributed his own success to the interest taken by his uncle in his early education. Not only did he learn about the natural sciences, he also acquired the skill to communicate his knowledge, and this proved vital to the Heath Society's program.

The links between Schouw and the leadership of the Heath Society were more than familial. Georg Morville had been a student of Schouw's and frequented his home. These two founders of the Heath Society, Morville and Dalgas, were thus in close contact with a man who argued for precisely the relationship between scientific research, general scientific education and social change which was to characterize the Heath Society's program. Schouw can be credited not only with laying the systematic foundations of plant geography and contributing significantly to geomorphology and climatology, but with helping to establish the foundations of modern geographical education through his internationally successful textbooks, which combined scientific method with a readable style. His educational efforts went beyond the schools, he was active in organizing lecture forums and popular journals for the purpose of spreading scientific knowledge. His lectures were eventually published in book form and, in translation, reached a world audience.

Schouw's interest in scientific education must be seen in the light of his advocacy of a democratic form of government. Democracy, he felt, required an educated general populace. Scientific education, particularly, gave an understanding of the process of growth and development of society in relationship to its environment, which would be necessary if Danish material and social life were to be improved. Understanding the changing character of the world was also politically important. People would not be moved to change their political institutions if they believed that the present system was part of an unchanging world order. This was an issue of actual concern to Schouw because he was actively engaged in politics, becoming a key figure in the peaceful movement to replace absolute monarchy with a democratically elected legislature under a constitutional monarchy in 1848–9. Therefore, whereas Mallet, the ideologist of absolutism, had argued that the Danish national character had been determined in an original wild state, Schouw argued against such environmental determination:

> can it be stated that the different character of nations is determined by, or at least essentially dependent upon, the nature which surrounds them? . . . Such a dependency is pretty generally assumed by historians, philosophers, naturalists and poets; but, nevertheless, I dare assert that this opinion represents a great error, which has only become so common because conclusions have been drawn upon the subject with a superficiality which would not be endured by any other science.[16]

Schouw saw the natural environment as mediated by human activities, rather than determining them. In this he was in accord with the classical conception of nature which saw man as playing an active role in the development of his environment:

> Sicily, formerly the granary of Italy, certainly produces much less corn now; many tracts of land lie desert, but this is to be ascribed to

> the deficiencies of social circumstances, not to the climate . . . If the social conditions of Algeria could be reduced to order, the fertility there would certainly not be inferior to what it was in antiquity.[17]

Like Steen Steensen Blicher, Schouw did not see nature as a cosmic principle, but as the creative, developmental potential within both humanity's material environs and humanity's physical being:

> Man is part of nature; she acts upon him, and he is subject to her laws; yet man stands, as it were, outside nature, and hence is capable of reacting upon her in a totally different way from all other creatures – of transforming her, and even to a certain extent to conquering and prescribing laws to her.[18]

In many respects the program of the Heath Society represented the realization of Schouw's ideals. It was a child of the new constitutional order in which reforms could no longer be promulgated from above as they had been under the absolute monarchy, but required broad public support. It was a program based on the belief that the spread of knowledge would pave the way for social improvements, and that these in turn would bring about the transformation of the environment.

Both Schouw and the Heath Society emphasized the importance of scientific education in bringing about environmental and social change. The arts, however, also played an important role in their thinking. Artists did much to stimulate the ideas of national regeneration and there were close ties between the Schouw and Dalgas families and the artistic community, including the authors Hans Christian Andersen and Adam Oelenschläger, the poet and educational/religious reformer N. F. S. Grundtvig and the sculptor Bertel Thorvaldsen. Dalgas's brother Carlo achieved recognition as a landscape painter and his son Ernesto gained entrance to the Danish pantheon as the author of soul-searching philosophical fiction. In his youth Ernesto travelled with his father inspecting the various activities of the Heath Society in the field, and endeavoring to inspire its work in verses such as the following:

> "Is there life in our Land?
> Are our thoughts on fire?
> Does a regal bearing inform each word and deed?
> Does the rumor fly abroad
> in foreign lands
> that Danish Viking spirit is reborn?"
>
> There *is* life in our land,
> It has yet to find its tongue
> We wait upon its birth. This is a time of Advent:

The grain bursts from the soil,
first in green, then in gold –
New shoots are visible each morning, if we care to look.

"Where is life in our Land?
Here is but death, here is but sand
Here are blowing dunes, and the Alheath soil.
Life will demand the years from us,
It will try us severely
Do we lack that will, to work and to believe?"

Our duty is to see if we cannot
bring that life to our land.
In digging death from the heath, we dig up treasure.
Quietly, humbly,
Death must be removed from the earth
on the darkest of nights, and before the cock crows.[19]

The promulgation of the Heath Society program

Dalgas sought to accomplish the Heath Society's goal of stimulating interest in the heath by publishing his *Geographical pictures from the heath* in 1868–9. It is regarded as a geographical "classic." Its publication represents a watershed in the Danish perception of Jutland because of its popular success and Dalgas's ability to counter established beliefs concerning the heath's infertility. The title might have been inspired by the critic P. L. Møller's statement that Blicher's "domain is first and foremost the Jutland nature, both that within the people, and that about them". This, Møller believed, was of potential political importance because "though a people may become quite apathetic, it will never go so far that they won't feel deeply moved . . . , when they see their most characteristic, half-forgotten national character captured and displayed in lively pictures".[20] Indeed, Dalgas begins *Geographical pictures* with a long passage from a story by Blicher depicting the changing landscape eastward along the road from Viborg to Randers:

> When one who travels from Viborg to Randers has passed the north end of the long lake which bears the city's name, and ascended the first steep heather hills, then a great heath spreads out before the traveller's eyes. It is called the Tap heath – a name which recalls memories, memories as dark and melancholic as the region itself. Two or three small huts, each surrounded by a little piece of cultivated soil, afford the only diversion on the two-mile long road to the country village of Ørum. From here on the road meanders between heaths and cultivated fields. Villages and churches arise up ever closer to each other,

> and the wide horizon is enlivened by distant forest. At the village of Leisten, one and a half miles from Randers, one says goodbye to the heather and forgets it instantly because of the delightful piece of landscape which seems suddenly to be conjured up from the last dark brown hills, and is presented to the traveller's delighted sight. Below is spread one of the country's largest fens and above it rise high, wave-formed, forest covered hills, enclosing a dale in whose background lies the castle named "Foussingø".[21]

Blicher's journey from wilderness to an enclosed idyllic rural landscape is intended, as the narrative later makes clear, to recall the golden age with its "great meadows enlivened by animals both tame and wild," a landscape which gives the "true appearance of paradisical peace and tranquility".[22] Dalgas's use of the passage suggests that the progress from wilderness to golden age landscape can occur in time as well as space. He writes:

> Blicher described this region as it was 30 years ago, and his description was not fiction. Even 15 years ago it was in all essentials true to reality. Now, however, conditions have changed. The heath between Ørum and Foussingø has largely disappeared; good fields with well built farms have replaced it, and prosperity has taken up residence.[23]

The landscape of the heath has changed over time; its nature is not fixed.

The developing character of the heath landscape is the central theme of *Geographical pictures*. Dalgas adopts the "bardic" pose atop an elevated barrow and describes the landscape vista that his gaze commands. It is these views, seen on a journey across the heath, that make up the book's geographical pictures. Like Blicher, Dalgas does not concentrate upon untouched sublime wilderness but rather he interprets the landscape as a material medium expressing the interaction of society and its environment over time:

> From the elevation of the barrows at Finderup, where we will take a little detour, we are able to see the whole of that sandy moraine belt which, in the wildest forms, creates something like a colossal beach running north and south which we may suppose had the function of protecting the morainic clay hills to the east from an imaginary ocean in the west, where the great heath flats now lie. Here and there from the barrows one sees low lying lakes, such as Rosborg Lake, known for its good duck hunting, and Bredsgaard Lake; the smoke from Finderup farm can be seen over the hilltops. Otherwise we have only wild heath before us, steep black hills in the most daring forms, and dark, deep dales. The flat Tap heath, whose heather cover may be a consequence of the decline of agriculture after the Swedish War, we

Figure 5.2 "Sketch of the heaths in Jutland, 1866," from Dalgas (1866). The encircled areas named "*slette flader*" are the "flats"; the open areas, "*slette bakker*," refer to areas with hills that are 50 per cent heath; and the dark north–south line of dashes identifies the "Jutland ridge". "*Rullestens Leer*" and "*Rullestens Sand*" refer to clay and sand moraines.

> would see disappear before the plow without complaint because, given its character, the region would be enlivened by cultivation. Here in Denmark's most characteristic heath district, it is not, however, without some sorrow that we reflect upon the fact that these proud heather hills also will disappear in time. Still, unless we close our eyes, every step we take will make us more and more aware that we are wandering over an ancient forest floor, and the requisites for afforestation are present here in full measure . . . It would not, of course, occur to us to take such a tour of the heath without taking along a spade and iron bore, and with their help we discover that the heath crust is evenly thick, and that hard pan is to be found underneath it. The subsoil is sand, but with enough clay to provide nourishment for deep tree roots. A few steps further, and we come upon white forest anemones, bilberry bushes, and, finally, oak brush: remains of the old oak forests. Yes, there is no doubt that this heather-grown sand ridge was once forest. Witness the one patch of oak brush after the next. The forest must, in its time, have stretched all the way from Hald Lake far to the north of where we now stand. It is clear that there was once a great forest here because we still can find colossal trunks of common forest pine and oak in the bogs. A few place names, such as Agerskov [forest field], Vedhoved [wood head], Skovsø [lake forest], Egebjerg [oak hill], and Skovhus [forest house], can still be found in the now treeless district, and stem evidently from a time when the forest had not yet disappeared. They bear witness against our forefathers, whose improvidence has deprived the country of such a climatically and economically important forest belt. They call upon us today to recover that which our fathers forfeited.[24]

Just as Blicher's landscape description was filled with symbolic meaning, so also were Dalgas's landscapes. Once again the landscape provides a veil through which to look at greater matters. The passage is replete with references paralleling past national crises with the present. The Tap heath, as we are told, is a product of a 17th-century war with Sweden which not only resulted in the definitive loss of the Danish provinces in what is now Southern Sweden, but also in repeating enemy occupation of Jutland. The wartime destruction of town, forest and field, and the decimation of the population by plague, resulted in the abandonment of extensive agriculture in many places, and the spread of heath. The "improvident" militarism of Denmark's forefathers had led to the loss of territory both within the kingdom and without. Finderup, where Dalgas stands astride the barrow graves of ancient heroes describing the landscape below, is a place well known in ballad and story, and loaded with symbols of earlier national crises. It was at Finderup that King Erik Klipping was murdered in an infamous regicide in 1286. The Danish noblemen who had forced him to sign a kind of Magna Carta then lost power to those who opposed such limitations upon royal power. This was regarded as one of the sordid events initiating a century of political dissolution which threatened the very existence of the kingdom. It was then, according to tradition, that the free farming class began to disappear as a political and economic force. It was also a time, too, of regional revolt, particularly in Jutland, of the Black Death and of abandonment of farm to heathland. Dalgas later claimed that:

> there can be no doubt but that about 15 000 square kilometers of our little country, and perhaps more, have been utterly destroyed in historical times. This is due, in part, to the terrible misfortunes which flowed over our country from the time of the Valdemars into the 17th century, and which were felt especially keenly in Jutland. It is also because the farmer class, here, as elsewhere, soon ceased to be land-owners, and because the nation's regents remained passive in the face of the troubles in the Jutland interior . . . .[25]

Dalgas's references to a past, now destroyed, national glory still evident in the landscape, recalls the arguments of the gothicists. The Tap heath with its memories, "dark and melancholic as the region itself," and its moral reminder of ancient deeds and of the need to make amends for those deeds, is used by Dalgas by way of a negative rural image to comment upon unnatural social behavior. More recent evidence of heath reclamation was, by contrast, an indication of society's return to a more normal state:

> But what is the reason for this healthy rate of development since 1848? In part, it is due to that life which was awakened in that year, and the improved social conditions which freedom brought with it . . . .[26]

Blicher had employed the condition of the environment, in classical fashion, as a means of expressing normative values concerning the state of the nation. He also, however, saw it as a particular landscape, a field of personal care and a material entity subject to study by agronomist and geographer. Dalgas follows in the footsteps of Blicher and Schouw by combining a literary perception of the landscape with a sophisticated scientific approach using the techniques of geographical science. With these methods he is able to demonstrate that it was not hardpan which had produced the heath. Rather, the removal of the forest cover had created the windswept conditions which favored the growth of heather, which in turn produced a leached, acidic soil and the hardpan in a process which was still continuing.

Dalgas used the natural sciences to show how social conditions influenced change in the physical environment. Science countered

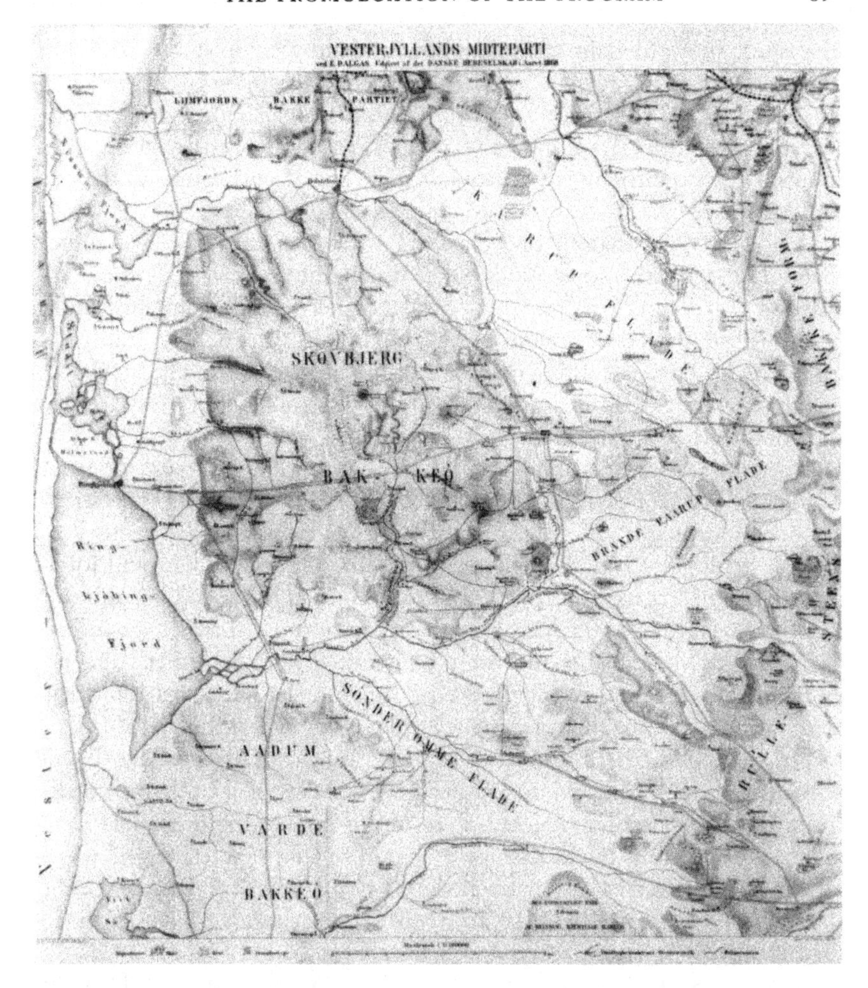

Figure 5.3 A map from *Geographical pictures from the heath* showing a variegated terrain of flats and "hill islands" (Dalgas 1866–7).

Forchhammer's "geognostic" conclusion that physical processes had determined the condition of the heathlands. The belief that the heaths were one vast plain underlain by hardpan was particularly damaging to the Heath Society's case because it created images of desert and sea, with their elemental monotony and association with savages and beasts. Blicher himself had used images of sea and desert and of destructive desert nomads. Dalgas too, employed the same imagery:

> in these desolate, vast deserts, medieval destroyers of peace had their haunts, and from here they practiced their extortion of the more prosperous districts.[27]

It was one thing to claim that areas of heathland were in such a state, but quite another to argue, as Forchhammer had done, that the entire region was one vast "hardpan flat". For this reason Dalgas had to demonstrate that Forchhammer "was not always in agreement with reality," even though this was difficult at a time when the word of a professor was held in rather greater esteem than that of an engineer.[28]

Dalgas showed that, if one departed from the normal travel routes through the heaths, one would discover large relatively fertile "hill islands" (*bakkeøer*), a word which has since become standard in Danish geomorphological terminology. These hill islands were demonstrably not underlain by hardpan and neither was much of the plains. The heaths, in fact, had a variegated landscape by no means destined by physical nature to the condition of a barren desert.

When Meir Goldschmidt met Dalgas on his heath journey, the latter was accompanied by Professor Johannes Johnstrup. Johnstrup was Forchhammer's protégé and successor. According to a standard history of Danish geology, "Dalgas's practically oriented heath studies, and Johnstrup's scientific geological studies, informed each other and brought the understanding of the genesis of Denmark an important step forward".[29] The reason is that Dalgas not only confronted Forchhammer's theory, proving the weakness of received wisdom, but also provided Johnstrup with information that helped him in the development of the first reasoned glacial theory for the origin of the Danish landscape. According to this still-current theory, the main terminal line of the last glaciation ran north–south down the center of Jutland, resulting in the desposition of rich, hilly, morainic soils to the east, and a flood of runoff water to the west, creating an outwash plain of leached sandy soils. The hill islands are the remnants of morainic deposits from earlier glaciations which survived the runoff.

Johnstrup's break with Forchhammer's diluvian theory and Dalgas's concern with the historicity of the landscape were complementary. As Clarence Glacken has written:

> The conviction that earth history had been more complex than the creation, the deluge, and the retirement of the waters, that the changes the earth has undergone since the creation are the results of natural processes . . . , that its history may be divided into epochs, suggested that man had also contributed to earth history.[30]

For Dalgas, as for Schouw, man was thus both part of nature, and a creature who stands outside nature. He bears a continuing responsibility for the state of his environment. This responsibility demands practical action:

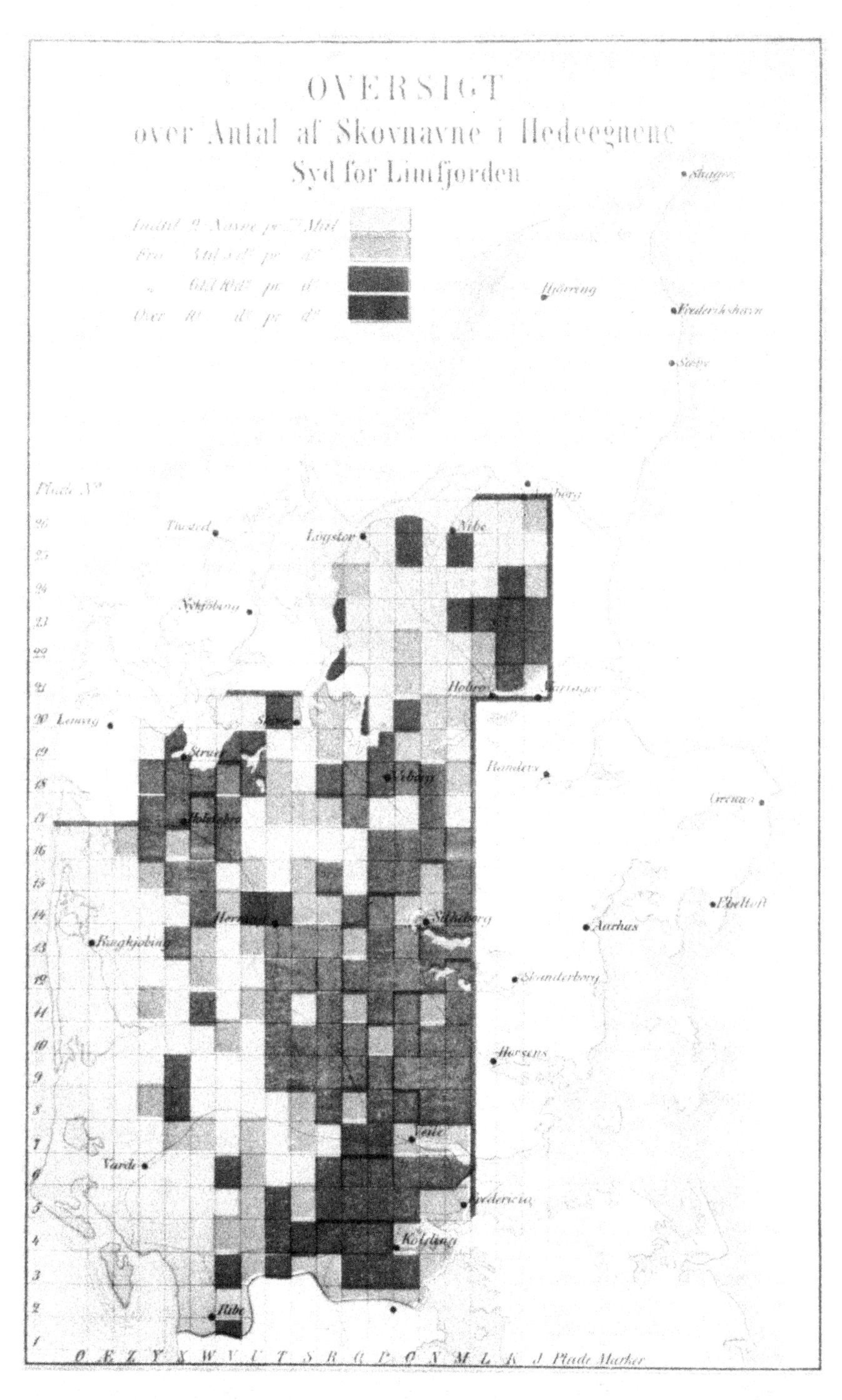

Figure 5.4 A choropleth map by Dalgas (1844) showing the number of place names indicating forested conditions per square mile.

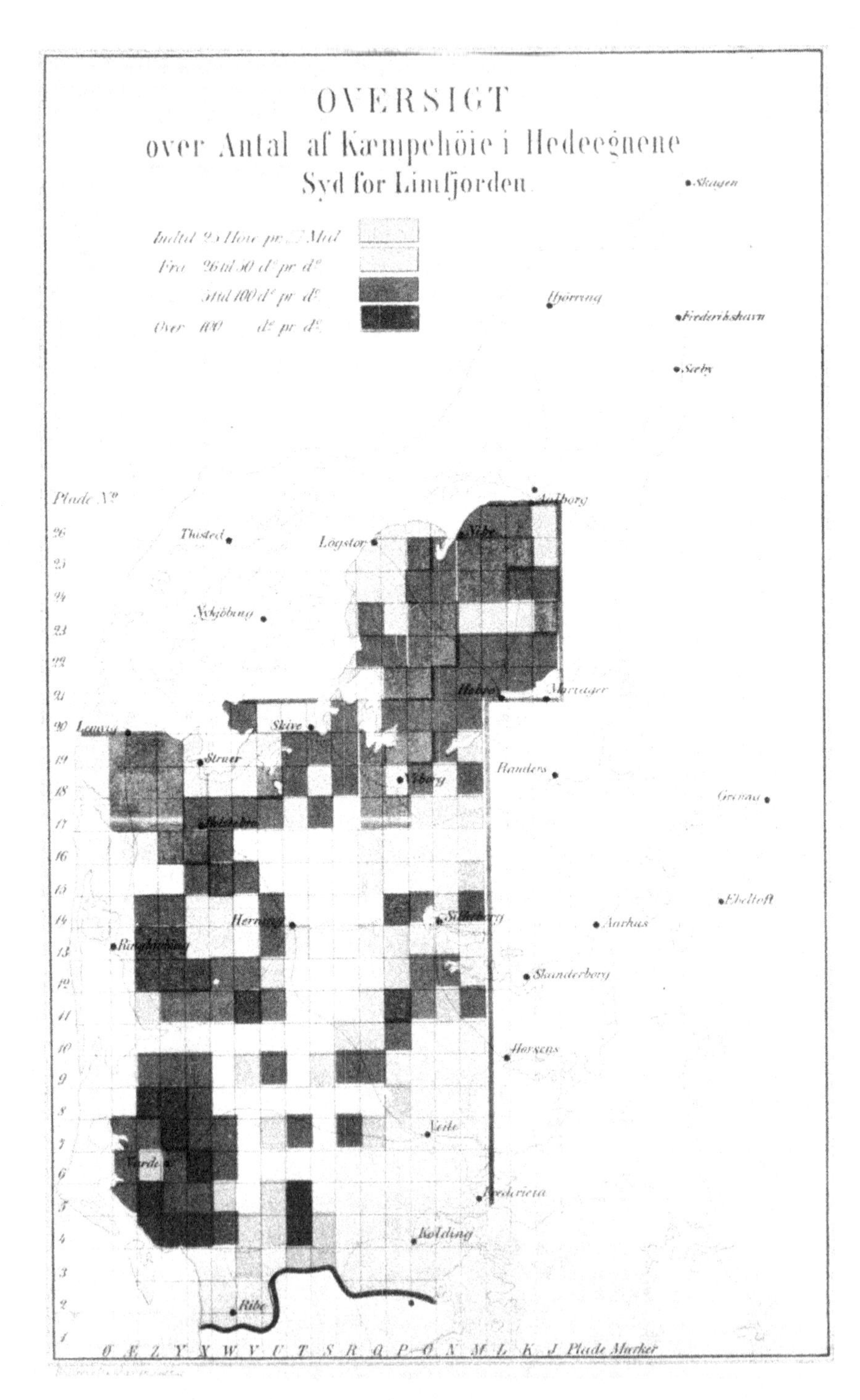

Figure 5.5 A choropleth map by Dalgas (1884) showing the number of barrows (an indication of early human settlement) per square mile. Figures 5.4 and 5.5 show that the earliest Jutland settlement was characterized by a landscape of forest intermixed with clearance, with a belt of forest down the center.

> Where the subsoil has the necessary properties, every decline in agriculture, every cutting down of forest, will call forth Jutland's greatest plague, hardpan. Many areas of sand, where the hardpan has yet to gain power, will, in time, be destroyed by it unless there is quick intervention with preventive precautions.[31]

The practical was coupled to an equally strong moral and social imperative:

> it must be the task of the present time to begin making amends for the sins of the past. It is only a question of strong will and faith: the Heath Cause is worth the sacrifice. Would that that will and that faith continue to grow.[32]

Dalgas's approach was to juxtapose a literary expression of social norms to scientific analysis of the physical environment. His landscape hides not only the graves of past heroes, but the buried ruins of ancient forests lying, as the passage describing the view from *Finderup* declares, along a ridge of hills whose north–south line separates heath flats to the west from more fertile soils to the east. This forest-covered ridge is given the character of a natural fortification for the ancient Danes. The defensive works of *Trælborg* thus:

> follow the crowns of the hills in such a way, that it is apparent that it has served to defend against an enemy coming from the west. That battles have occurred here would seem to be indicated by the name of the village which lies behind the embankment: *Leirskov* [camp forest], and from the many great warrior barrows which lie about the area.[33]

Names like Finderup would bring to his reader's mind ancient periods of national decline, whereas Dalgas's presentation of the forest as an ancient line of fortification would recall the recent fall of Dannevirke. Like Dannevirke these natural embankments had been neglected, but it was not too late to:

> reinstate these earthworks to their former dignity as barricades for Jutland's eastern half. In a prehistoric period they might have protected it as a beach against the North Sea; later, they were covered with forest, and protected against the west wind; but now they are left deserted and without use.[34]

The forest had provided shelter from wind and storm for growing crops. When the forest was felled only heath vegetation could survive.

Dalgas's identification of trees with fertility is commensurate with the landscape ideal of the gothicists and the rural idyll of literature. Trees were not only to be planted in a great belt protecting eastern Denmark, but as shelter for field and gardens. They would thereby restore the balance of the elements and return fertility to the land:

> gardens will have great importance for Jutland because it is primarily through them that a desire for the planting of shelterbelts and trees will be awakened. This in turn, when accomplished on a large scale, will have an important climatic influence on our country . . . If Jutland were afforested, as it ought to be, I venture that the country's rainfall would undoubtedly be somewhat more favorable to agriculture than it is now.

The images of a rural idyll which inform so much of Dalgas's thought occasionally become explicit:

> Gardens exercise a wakening, beneficial, elevating and softening influence on the mind, and they provide support for a lovely life together in the home.[35]

Dalgas takes up the Virgilian theme – in which plowshares replace swords – in the martial images used to describe the peaceful "battle," behind bulwarks of trees, against the hardpan. His conclusion to *Geographical pictures* is characteristic. It is a call to return to the forest its "lost dominions" and a wish that:

> these Heath Pictures [may] bring you the conviction that the Heath Cause is of very great importance to the fatherland, and that the difficulties which will be met can be overcome . . . and that it should be every Danish man's duty to join in removing the curse which seems to have lain upon West Jutland for a half-millennium, a curse which, among other things, has resulted in the complete destruction of the forest.
>
> 1200 men now comprise a small but close-knit phalanx, the Danish Heath Society. They have already dug the trenches in pursuit of that objective which must be gained. Much greater support is absolutely necessary.[36]

Dalgas's efforts as a publicist were crowned with remarkable success. The leader of the opposition within the Royal Agricultural Society, la Cour, capitulated in the face of what he saw as massive public sympathy for the Heath Society, and even took an active part in its work. The Heath Society mobilized both public and private support and, with it, capital for its projects. It began works of drainage, irrigation and afforestation. Of greater importance, however, was the myriad of private and local works

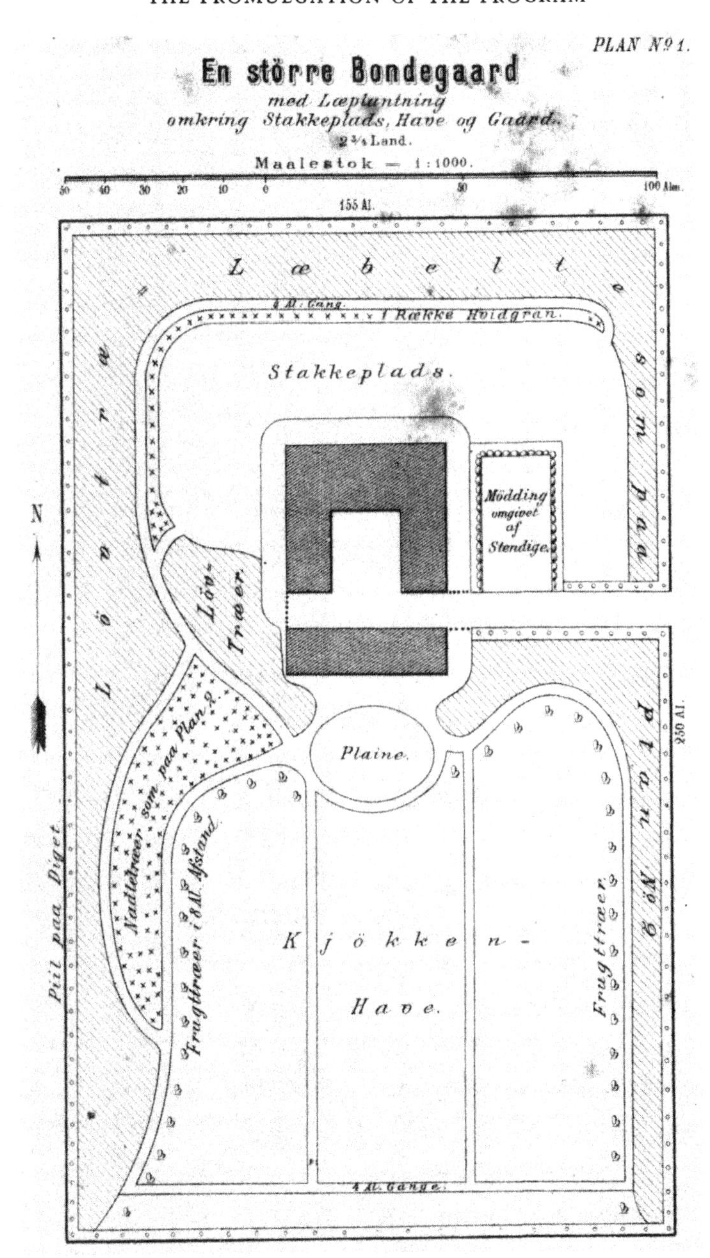

Figure 5.6 A plan for a farm garden by Dalgas (1875). The diagonal hatching indicates a shelter belt of trees. There is a lawn, an orchard and vegetable garden.

which it inspired by turning heath reclamation into a national patriotic cause. Capital poured into state and private afforestation. The returns soon flowed into the pockets of local farmers in the form of off-season employ-

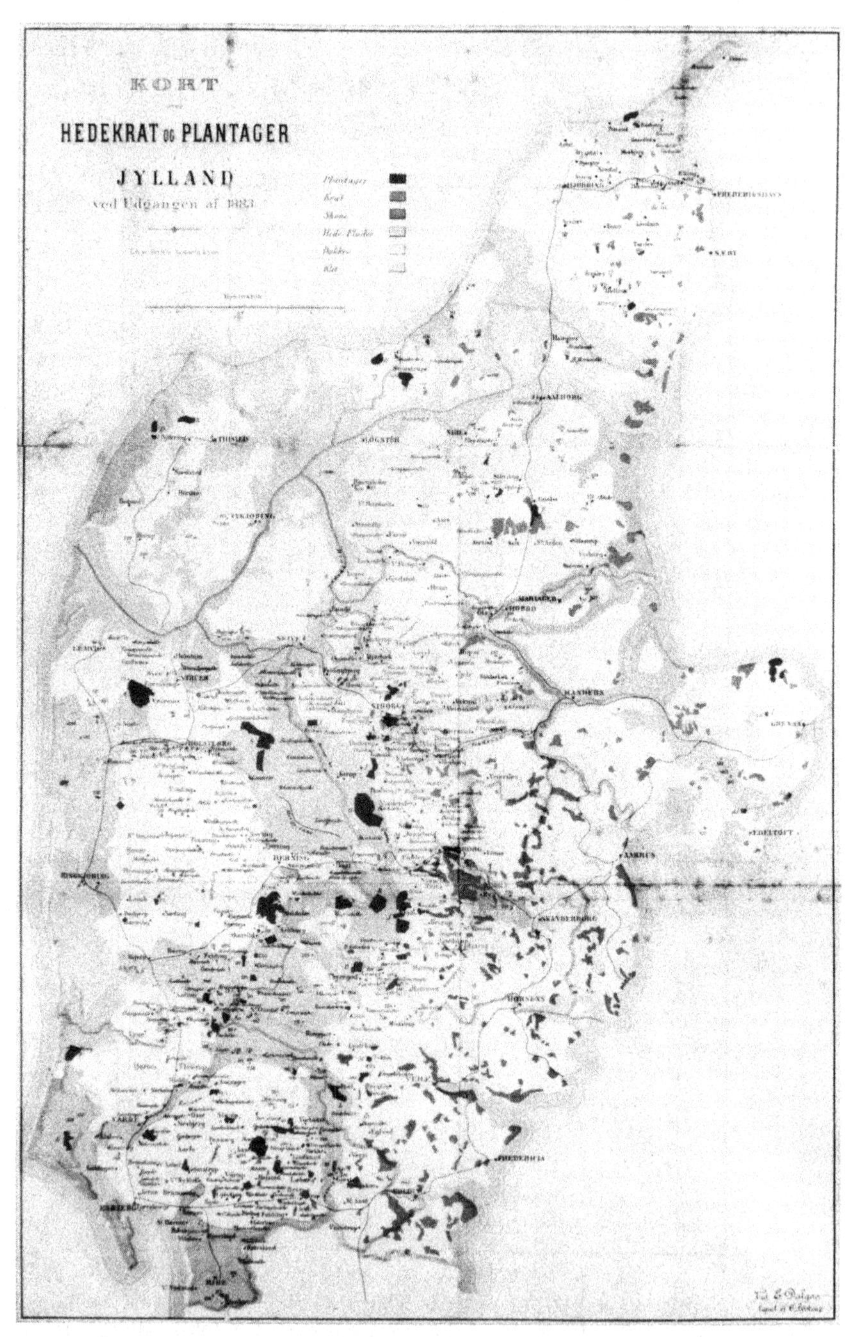

Figure 5.7 An 1883 map by Dalgas (1884) showing areas of new afforestation (*plantager*), brush (*krat*), established forest (*skov*), heath flats (*hede flader*), hill islands (*bakkeøer*), and dunes (*klitter*).

Figure 5.8 An afforested area on the Alheath.

ment in the forests. With this money, often guided by the Heath Society, they were able to make agricultural improvements and intensify production.[37] Infrastructural developments had similar multiplier effects and facilitated the sale of agricultural products. The area of heath declined from about one million acres (400 000 ha) in the early 19th century to about 500 000 acres (200 000 ha) in 1907, or from about 40 percent to 19 percent of the area of western Jutland including Viborg, Ringkøbing and Ribe counties. By 1961 the total had dropped to approximately 200 000 acres (81 000 ha). For Jutland as a whole the figure dwindled from 1.6 million acres (600 000 ha) to 300 000 (120 000 ha) in 1961. In the same time period the percentage of forest in Jutland increased from about 2 percent to 40 percent and over 15 000 km of shelter belts were planted.

Gradually the Jutland landscape took on the enclosed, forested character that supposedly had existed in ancient times, and the population too returned to supposed former levels. In Ringkøbing County, a typical heath district, the population rose from 55 600 in 1840 to 164 500 in 1940, the result of internal population growth and immigration from adjacent, non-heath areas. Agricultural intensification and specialization together with a growing population provided the basis for industrial development in textiles, food processing and farm service. Urban development followed, taking Jutland's urban : rural ratio to that of the rest of Denmark by the mid-20th century.[38]

Dalgas was regarded as "the hero of the heaths".[39] After his death in 1894 a suitably "ancient" memorial was erected to his memory. It consis-

Figure 5.9 A present-day view from a barrow at Finderup in the snow. The cultivated field is bound by shelter belts, and a small patch of heath is visible on the right.

Figure 5.10 The garden and farm dwelling at Ole Vestergaard Jensen's farm, Drongstrup, Sønder Felding. The bridge is over an irrigation canal, planned by Dalgas, which taps the Skjern River.

Figure 5.11 The 1894 Dalgas memorial.

ted of a stylized barrow of stones or monoliths. In the 1950s a memorial heath park on the remains of the Alheath was dedicated. In an enclave in the heath a circle of stones has been placed in honor of individual heroes of the heath cause. They include both Blicher and Dalgas. Stones for the collective heroes of each heathland district have also been erected in two parallel rows leading to the enclave. On each stone is chiselled a pie diagram showing the percentage of heath, forest and cultivated land to be found in 1850 and 1950 in the heath district which it represents.[40] Dalgas himself, however, would probably be best pleased with another monument. It consists of a statue of the reclaimer standing upon an ancient barrow and spying out over what once was a desolate heath-clad landscape. It overlooks a now cultivated and forested scene. Gothic Jutland has been realized!

Notes

1 For more detailed background on this period, English language readers are referred to Oakley (1972), *A short history of Denmark*, pp. 180–205.

(a)

(b)

Figure 5.12 The memorial circle at Kongenshus Memorial Park on the Alheath. (b) Stone for the district of Lysgaard with verse by Blicher.

Figure 5.13 Dalgas and friend spy out over the former heaths.

Figure 5.14 A portrait of Dalgas (photo courtesy of the Royal Library, Copenhagen).

2 For biographical background on Enrico Dalgas, see: Dalgas (1891), *Familien Dalgas: Slægtsregister fra 1685 til 1891 med Beretninger om Familiens Medlemmer og mine egne Livserindringer*; Skodshøj (1966), *E. M. Dalgas*; and Skrubbeltrang (1966), *Det Indvundne Danmark*, pp. 203–321.

3 Dalgas (1865), Om Opdyrkning af de Jydske Heder: in *Berlingske Politiske og Avertissements-Tidende*, no. 287 (Wednesday, November 15) and no. 288 (Thursday, November 16).

4 Dalgas (1891), p. 56.

5 Skrubbeltrang (1966), p. 130.

6 The lines are by the poet H. P. Holst, an ardent patriot with a military background. The lines appeared on a medallion produced for the Nordic Exhibition in Copenhagen in 1872 (see Skrubbeltrang 1966, pp. 116–17).

7 Dalgas (1887), *Hedesagens Fremgang: 1866—87*, p. 7.

8 Quoted in Skodshøj (1966), p. 90.

9 For la Cour's views, see: la Cour (1865), Landbruget i Kempen (Campinen), pp. 217–85; la Cour (1866), Om Hedernes Benyttelse, pp. 141–6; and Dessau, D. 1866. *Den tiende danske landmandsforsamling i Aarhus 25—29, Juni 1866*, pp. 195–206, 216–17.

10 Oppermann (1889), *Bidrag til det danske Skovbrugs Historie 1786—1889*, pp. 242–4.

11 Forchhammer (1835), *Danmarks geognostiske Forhold, forsaavidt som de ere afhængige af Dannelser, der ere sluttede, fremstillede i et Indbydelsesskrift til Reformationsfesten den 14de Novbr. 1835*, p. 105.

12 Forchhammer (1844), Om en stor Vandflod, der har truffet Danmark i en meget gammel Tid, pp. 95–6.

13 Though Forchhammer's work provided ammunition for opponents of the Heath Society's program, Forchhammer himself believed that it would be possible to cultivate, if not afforest, the heaths (see: Forchhammer (1855), De Jydske Heder, pp. 169–70).

14 Olwig (1980), Historical geography and the society/nature "problematic": the perspective of J. F. Schouw, G. P. Marsh and E. Reclus, pp. 29–45.

15 Dalgas (1891), p. 29.

16 Schouw (1852b), Nature and Nations: in *The Earth, plants and man*, p. 241.

17 Schouw (1852a), The action of the human race upon nature, p. 238.

18 Ibid., p. 228. On Schouw's ideas about man and nature, see Olwig (1980).

19 On the ties between the heath cause and the arts, see Erichsen (1903). *Den Jydske Hede Før og Nu*. The poem was published posthumously in Dalgas (1903), *Sangbog* (Axel Mielche, ed.), pp. 90–1.

20 Garboe (1961), *Geologiens Historie i Danmark*, vol. II, pp. 332–8; see also Oppermann (1889), p. 260.

21 Møller (1971, orig. 1847), S. S. Blicher, pp. 176–7; Dalgas (1867–8), *Geographiske Billeder fra Heden*, p. 3.

22 Blicher (1927), Eneboeren paa Bolbjerg: in *Samlede Skrifter* (J. Aakjær & G. Christensen, eds), vol. XIX, pp. 123–4.

23 Dalgas (1867–8), pp. 3–4.

24 Ibid., pp. 12–13.

25 Dalgas (1884), Fortids- og Fremtids- Skovene i Jyllands Hedeegne: in *Hedeselskabets Tidsskrift* V, no. 1 (Jan., Feb., Mar.), p. 71.

26 Dalgas (1867–8), p. 6.

27 Ibid., pp. 103–4. See also Erichsen (1903), p. 238.

28 Ibid., vol. II, pp. i–viii.

29 Garboe (1961), p. 336.

30 Glacken (1967), *Traces on the Rhodian shore*, pp. 704–5.

31 Dalgas (1867–8, p. 39) held the now accepted theory that hardpan formation was the result of water percolation through the acidic layer of humus built up by the heather; this acidified water then was able to dissolve iron and minerals from the soil layer immediately below the humus, redepositing them as hardpan further below.

32 Ibid., p. 44.
33 Ibid., pp. 22–3.
34 Ibid., p. 17.
35 Dalgas (1875), *Anvisning til Anlæg af Smaaplantninger omkring Gaarde og Haver samt til Anlæg af levende Hegn og Anlæg af Pileculturer*, pp. 3–4. Dalgas was a founder of the Jutland Garden Society – see Dalgas (1873), Det Jydske Haveselskab, pp. 366–9.
36 Dalgas (1867–8), p. 125.
37 Grøn (1933), *Hvad Nytte er Skoven Til?*
38 Useful summaries of the impact of the heath reclamation movement can be found in Skrubbeltrang (1966) and Struckmann *et al.* (1943), *De Danske Heder: Deres Natur og Fortidsminder, Folkeliv og Kultur*, 2 vols. For geographical data, see especially Aagesen (1943), Fra Hedegaard til Bysamfund: in *De Danske Heder*, vol. II (Struckmann *et al.*, eds), pp. 263–98.
39 Riis (1910), *Hero tales of the far north*, pp. 153–77.
40 Skodshøj (1953), *Hedens Opdyrkning i Danmark*.

6 *Divergent views of the heath*

The introduction of full parliamentary democracy in Denmark in 1901 ended the country's long recurrent political crises by fully investing power in the legislature. This had been the aim of the liberal Farmers' party, which had dominated parliament, but their control of the new system was threatened by the changing character of Danish politics. Denmark was now a rapidly industrializing society. Its cities were growing, the rural landscape was undergoing transformation. Dairies and slaughterhouses producing for export were increasingly common. Denmark met the threat of cheap grain from Eastern Europe and America by shifting production to high quality meat and dairy products for export to nearby industrial markets, particularly England. The transformation of agriculture and its associated industries produced rural and urban proletariats. They found effective political leadership in the Social Democratic party which challenged the political hegemony of the farmers. This change in Danish society was reflected in a changing view of the nature of the Jutland heath. The heath had been seen as a symbol of the potential for development buried in the Danish soil and its people. Now that this potential had been realized and the Heath Society had become a national institution, heath cultivation likewise became institutionalized as a cause for itself, outside the wider political and social perspectives of Dalgas's era. It was at this time that a movement to preserve the remaining heaths developed, a movement which questioned the social, esthetic and ecological basis for continued heath cultivation.

Jeppe Aakjær

The central figure in the changing perception of the heath in the early years of the 20th century was Jeppe Aakjær (1866–1930). Born Jeppe Jensen, he took his pen-name, Aakjær, from a little hamlet on the Karup river south of Skive where he had grown up. His family had a poor farm in a heath area, and Aakjær identified strongly with both his home district and the rural poor. He is best remembered as a lyricist of the heath, the man who followed Blicher as poet and realist writer.[1] Aakjær was also an active public speaker and journalist, a Social Democrat who agitated for social reform both locally and nationally.

It is unusual for an author to immerse himself quite so deliberately and methodically in the life and work of a predecessor as Aakjær did in the case of Blicher. He became Blicher's literary executor, publishing a three-volume biography and compiling and editing the definitive 33-volume edition of his collected works. By publishing Blicher's works in consecutive order, with political tracts and agronomic treatises alongside lyric poetry, Aakjær provoked the literary establishment by making an issue of Blicher's extra-literary interests. Without being derivative, Aakjær followed Blicher's example in his own poetry and the realism of his prose. He published scholarly works about his native region and wrote poetry in its dialect while cultivating an interest in British counterparts, particularly Robert Burns. His "Jenle" festivals, which took their name from his farm, were clearly inspired by Blicher's "Himmelbjerg" festivals, whose history Aakjær had studied in detail. But the thrust of Aakjær's interest in the heath was different from Blicher's. Aakjær was not interested in developing whatever potential the heath might have, but rather in preserving it from development.

The Heath Society at the turn of the century was very different from its early days as a kind of grass roots organization fighting for a cause. It was now a quasi-governmental agency, and had become an accepted patriotic commonplace, an unquestioned good cause.[2] This status was achieved at the very time that the *raison d'être* of the Heath Society – the unreclaimed heath – was beginning to disappear. The heaths that were most suitable for reclamation were either cultivated or afforested, leaving only the more marginal areas untouched. With this success the thrust of the Heath Society's efforts began to change. Technical and economic rather than social concerns came to the forefront. Emphasis was placed upon the supposed climatic improvement and economic gains of reclamation.

Dalgas had seen the heath cause as a means of supporting the native society in its efforts to improve the standard of living. In an early confrontation with the Royal Agricultural Society he had written:

> I must draw attention to the fact that Denmark is not made up of Zealand, Funen and East Jutland alone. There is also something which is called West Jutland, where many people live; they also farm, and their agriculture needs to be intensified. This can best be achieved by the means suggested by the Heath Society: afforestation and the creation of artificial meadows. It is not only for the sake of the dead earth that we have come here, but also for the living people, who also have a claim to make . . . .[3]

By the turn of the century things had changed. Heathland agriculture had been intensified and the condition of the farmers had improved. The land which remained, even by Dalgas's calculations, was of marginal value for reclamation. Aakjær argued that under these conditions heath reclamation had become a self-sustaining cause, which neglected the needs of the native population precisely for the sake of the "dead earth". The Heath Society had become, in Aakjær's view:

> a gigantic enterprise, an enterprise under the sign of the crassest materialism, and its cold rectangular principles move like an iron roller over this beautiful and unique landscape. It is a large and extraordinary enterprise which has always had the whistling wind in its sails.[4]

Aakjær felt that the fact that heath reclamation had become a self-justifying cause meant that it distracted attention from the causes of social problems by reducing them to a question of agricultural intensification. Heath reclamation had become a politically neutral panacea for a variety of social ills ranging from rural poverty to urban unemployment. Aakjær, who knew the heathlands well, was aware that the marginal soils which remained could not provide a solution to the problem of the landless rural poor. They needed good soil if they were to survive in a mechanizing rural world. Overcrowded cities hardly offered an alternative. To give them a patch of heath to reclaim was little more than a bitter joke upon these poor people, a sop, to which they would devote their lives and receive in return a worthless farm. He concluded that:

> The Heath Society, on the whole, has been a stick in the spokes of the wheel of progress, a conscious and active reaction against major land reform. It has centered debate on heather and sand, when it ought to have been concerned with those good Danish loams which the nobility have surrounded with barbed wire.[5]

The hero of Aakjær's novel, *The children of rage* (1904), characteristically prefers emigration to America over the prospect of creating a farm on marginal heathland.[6] For him, what is lost within must be won without. In Aakjær's opinion Jutland was a defeated country:

> Those who describe the Heath Society as a conqueror are correct: West Jutland is, in my eyes, a conquered country. Any Jutlander who thinks about what has occurred must regard the Heath Society and its so-called cultivation of our forefather's land with the same eyes as a conquered people looks upon the monuments to victory which the enemy raises upon the land of the conquered.[7]

The heath preservation cause

In his campaign against the reclamation cause Aakjær reinterpreted the literary conventions of the rural image. In *The children of rage* he creates a sense of outrage by juxtaposing expected images of rural tranquillity with the bitter reality of the rural poor. The novel ends like Virgil's first *Eclogue* with the forced exodus of the native from his land: "thus fares the land, by luxury betray'd". Aakjær reversed Blicher's image of the heath. It is no longer the ruin of a former rural paradise, but a pastoral paradise itself, and it is threatened by the unnatural activities of the Heath Society:

> no conquering horde could have gone more ruthlessly to work against a province's original national aesthetic values. It is a Society which threatens all that we hold holy, all to which we have declared our love in our life's most precious moments: the landscape, the view, the mile vaulted horizons. What a fairytale land these West Jutland districts were in our childhood, when sheepbells rang through the bedewed quiet of the morning. Miles wide over heather and moss the horizon lay in the misty distance, unbroken at the edges, while millions of spiderwebs shone in the bog myrtle encircling the smoking heath lake.[8]

Aakjær directly challenged the gothicist and national romantic perception of the heath as the landscape ruin of an earlier golden age, which the nation was morally obliged to restore:

> The Heath Society was a child of romanticism. Just as Oehlenschläger wished to raise the ancient spirit in the Danish people, Dalgas fantasized about raising the ancient forest on the Jutland heath.[9]

Aakjær preferred to see the remaining heath preserved as a reminder of the past.

> The Jutland heath, in my eyes, is the most characteristic, and undeniably the oldest part of the Danish landscape map. Let this shadow of ancient times continue to rest on our mother's weatherbitten cheek.[10]

There was ample precedence for Aakjær's approach to the landscape in that aspect of the pastoral tradition known as elegy. Rather than dream of restoring the golden age, elegy dwells with nostalgic sorrow upon the memory of a lost past. Blicher exemplified this in an early poem called "Homesick":

> My birthplace is the heather's brown land,
> My childhood's sun has smiled on dark heath,
> My infant foot has trod the golden sand,
> Among black barrows lives the joy of my youth.

> Beautiful to me is the flowerless meadow,
> My brown heath is an Eden's garden:
> There too my bones will rest some day,
> Among the heather-grown graves of my forefathers.[11]

The elegy not only provided Aakjær with an indirect means of criticizing reclamation that destroyed the signs of a former paradise, it also provided an argument for preserving the heath as a landscape memorial. Aakjær pleaded:

> Spare this remainder of seasoned ancient earth. The sunset bridge, the brown course of the wind, where ancient paths wind hidden in the heather, and the sky displays its bright banner. Do not disparage the heath's "barren" gift; it is a poor country which is nothing but garden.[12]

In appealing for the lost paradise of his own and the nation's youth Aakjær developed an argument for preservation as memorialization. His use of the elegy is in agreement with the classical conception of nature because the heath's physical environment symbolizes the natal state of the nation. The classical conception involved development toward an increasingly reciprocal use of the environment and was opposed to extractive uses. This meant that a reversion to pastoralism was *unnatural*, but pastoralism combined with an agriculture which used animal fertilizer to restore the soil's nutrients was in harmony with the classical ideal. The intensification of agriculture according to the "cold rectangular principles" of maximum economic gain threatened, in Aakjær's view, the marginal landscapes of heath and meadow. If Aakjær was the poet of the heath, he was even more the poet of the riverine meadow, which is what an "aa" "kjær," roughly translated, literally is. At the same time as agricultural intensification was removing these former open lands from common use, industrialization was creating large urban populations. To preserve such common land either for the regeneration of the soil or the regeneration, through recreation, of men accords with the spirit of the classical conception of nature.[13] The Social Democrats, together with the Nature Preservation Society, supported nature preservation in the interest of opening up the landscape for the recreative use and esthetic pleasure of the general populace. In their view nature was the common birthright of the nation. The heaths, and meadows, the forests and beaches should be accessible to all. As a Social Democrat, Aakjær supported preservation on principle and his poetic depiction of the esthetic and recreative pleasures to be found in the heath landscape provided a source of inspiration for the preservation movement.[14]

There was a small, but influential wing of the nature preservation movement which drew its members largely from the natural sciences and which wished to preserve nature *from* man rather than *for* man.[15] In his polemics against reclamation Aakjær cited these scientists, who regarded the heath as being "primordial," and he himself occasionally argued for the preservation of the heath as "wild," "untouched," nature.[16] The heath becomes, in this context, a place where "a soul can be alone with the Almighty, alone with the desert waste and the stars".[17]

In pursuing his cause against the Heath Society Aakjær did not pause to reflect upon the internal contradictions of his argument. The heath could not be both empty of people and yet a pastoral paradise where "sheepbells rang through the bedewed quiet of the morning". It could not be both a park for vacationing urban dwellers and a reservation where scientists could study undisturbed nature.[18] It could not be populated by grazing sheep and ancient monuments and yet be primordial and untouched wild nature. Esthetically, it could not be replete with human meaning through childhood associations while being an empty wilderness cathedral where nature exerts a direct influence on the spirit and one comes face to face with the Almighty. By the same token the nature which preserves an irregular landscape of memory was potentially in conflict with the landscape of rational scientific laws that could be applied equally as well, or better, to a scientifically organized forest plantation as to an irregular wilderness.[19]

Despite internal contradictions the preservation movement was able to unite to save the heath. Some areas were preserved for scientific reasons, others for historical and esthetic reasons. Much of the remaining 230 000 acres (93 000 ha) of heath is under preservation, and areas of preserved heath now dot the map of Jutland. At Kongenshus on the Alheath near Viborg 3335 acres (1350 ha) of unreclaimed heath have been preserved, while 2964 acres (1200 ha) remain at Hjerl heath near Skive and 4520 acres (1830 ha) of Borris heath are preserved near Skjern – to indicate the extensiveness of just a few of the preserves.[20] The poet's heath and the scientist's were apparently the same, as the prominent botanist Christian Raunkiær showed when he published a book-length analysis of the flora that appears in the poetry of Blicher and Aakjær![21]

Aakjær's varying conceptions of the heath's nature were paralleled in the contradictions of the preservation movement of his time. They remain codified in contemporary Danish preservation laws which classify as nature everything from ancient monuments to landscapes of historical, educational or recreational importance and sites of special scientific interest. The distinctions between these different "natures" is ignored in practice.[22] In consequence much of preservation effort is vitiated by natural processes altering the character of the legislatively "preserved" nature.

The "heath" today

The map of Jutland is dotted with areas of heath preservation, but the traveler may have difficulty finding the sort of landscape described by either Blicher or Aakjær. The heath of their days has largely disappeared, ironically, because of the efforts of preservationists. Compromise was reached among preservationists by taking their point of departure in their common conception of nature as a landscape to be protected from development. Preservationists did not take into consideration the fact that in reality the heath was a product of continuous interaction between society and environment. One could preserve a heath for science, for recreation, or as an historical relic, but without the periodic burning and grazing practiced by traditional heathland farmers the heather dies out and is replaced by shrub and tree growth. The heath is not wild nature, but a cultural landscape developed by farmers for use in a particular type of agricultural system.

Today the same cover of shrub and trees which is obliterating the heath is also advancing into the reclaimed agricultural land of the former heath. Concentration of landownership, mechanization of farming and the capital it requires demand a high rate of profit. All tend to favor more extensive land uses for marginal agricultural soils such as forestry, recreation or even abandonment. Most of the loss of cultivated land has occurred in a north–south belt paralleling the main terminal line of the last glaciation. In a 1900 square kilometer area stretching from Bække in the south to Silkeborg and the Karup river valley in the north, 60 square kilometers were abandoned as farmland between the turn of the century and 1950. In the period roughly between 1955 and 1965 alone a further 24 square kilometers were lost.[23]

Today, considerable efforts are being made to find ways to restore the heaths to their former bare state. Experiments include grazing, burning and the physical removal of unwanted vegetation.[24] Attempts have also been made to halt the abandonment of farmland. In 1967 a law was passed prohibiting the use of farmland for purposes other than cultivation without government approval. By their nature such laws are difficult to enforce and as a consequence the law was already liberalized by 1973.[25]

The attempts to preserve the landscape *status quo* suggest a continued concentration upon nature as a physical object rather than upon those social forces which shape it. The resultant failure to preserve the heaths and cultivated landscapes, and the very cost of artificial methods of doing so, raise serious questions about underlying contemporary conceptions of nature. How can nature be defined as primordial, the antithesis of the man-made, needing protection from man, and yet require human intervention to remove the vegetation that grows when it is thus protected? If, on the other hand, the heath is not itself nature, should one then allow nature to cover the barrows and other ancient monuments which are also preserved in the landscape of the bare heath? Again, if the heath is regarded as both nature and the creation of former agricultural practices, how does it differ from other agrarian landscapes that one might wish to preserve? Questions such as these direct attention away from the material entity toward the social values expressed in the landscape. Preservation then raises moral questions which point to the changing character of farm life and agriculture's use of the environment, rather than solely to esthetic or scientific issues of the value of given landscapes. Social and moral questions threaten the consensual neutrality of the preservation movement and are often avoided.[26] Meanwhile the heaths and other former common lands continue to revert to shrub and woodland.

The geography of the heath

We began with a passage from a standard geography of Scandinavia. In this passage the recent history of the heath was summarized as follows:

> The demand for new land rose especially after the loss of North Slesvig in 1864. Bravely and industriously, the local population waged war on the heather, encouraged after 1866 by the Danish Heath Society (Det danske Hedeselskab) . . . Only a few patches of heath remain, as scar-like reminders of the former waste, but in several places small areas of heath have been carefully preserved lest future generations forget the labours of the past.[27]

The passage gives the impression that reclamation was largely a response to demographic and economic necessities while preservation memorializes that effort. Its language, however, is loaded with the rhetoric of the Heath Society. That rhetoric conveys the spirit of a nationalist ideology aligned to practical farming needs. But the claim is unexamined, it is lifted uncritically from the publications of the Heath Society. It fails to ask why Danes waged war against a peaceful landscape which, had it been in Britain, for example, might have been carefully tended in the interest of grouse hunters or shepherds.[28] The real significance of the loss of North Slesvig is left unstated. Dalgas himself recognized that reclamation had been going on long before 1864. The war against Prussia did not create a significant increase in Jutland's population or a significant migration there from Slesvig-Holsten, and the opposition of the agricultural establishment to reclamation at the time suggests that the economic motive was weak. The motives behind the national movement to reclaim the heaths, and the

(a)

(b)

Figure 6.1 (a) A heathland barrow disappearing beneath a cover of shrub and trees; (b) heath in the process of being covered by shrub.

form which that movement took can only be fully understood in the light of a national ideology which consciously or unconsciously came to be identified with the heath. The same applies to the later movement to preserve the heaths. It is misleading to imply, as the textbook does, that such national movements are essentially rational responses to physical needs created by war and the desire to preserve the memory of past efforts.

To achieve a full understanding of environmental use and environmental preservation we need to examine the ideological use of nature and landscape to express ideas concerning the character of society and its relationship to its environment. This does not mean that ideology determines development. Heath reclamation and preservation took place within social, economic and environmental parameters that determined the practicality of the measures promoted. Those measures, however, reflected not only practical problems, but key issues, tied to the Danes' conception of the ideal nature of their society and the prospects for its future.

Notes

1 Nørgaard (1941), *Jeppe Aakjær. En Introduktion til Hans Forfatterskab*; Schmidt (1933), *Jeppe Aakjær, Nogle Oplysninger om Hans Forfatterskab.*
2 The post-Dalgas history of the Heath Society is detailed in: Geckler (1982), *Hvad Indad Tabes: Hedeselskabets Virksomhed, Magt og Position*; Pedersen (1971), *Hedesagen under Forvandling: Det Danske Hedeselskabs Historie 1914—1966*; and Skrubbeltrang (1966), *Det Indvundne Danmark*, pp. 325–469.
3 Dessau (1866), *Den Tiende Danske Landmandsforsamling i Aarhus 25—29, Juni 1866*, p. 210.
4 Aakjær (1915), *Hedevandringer*, p. 95.
5 Aakjær (1919c), Naturfredning (orig. 1909): in *Samlede Værker*, vol. IV, p. 477.
6 Aakjær (1919d), Vredens Børn (orig. 1904): in *Værker*, vol. VI, pp. 439–650.
7 Aakjær (1919b), Hedens Fredning (orig. 1916): in *Værker*, vol. IV, p. 488.
8 Aakjær (1919c), p. 476.
9 Aakjær (1919b), p. 486.
10 Ibid., p. 490.
11 Blicher (1920c). Hiemvee: in *Samlede Skrifter* (J. Aakjær & H. Ussing, eds), p. 137.
12 Aakjær (1918), Paa Hedens Høje (orig. 1906): in *Værker*, vol. I, p. 214. On Aakjær's pose as a shepherd, see Brostrøm (1969), Digt og Digterjeg, Jeppe Aakjær: Sundt Blod: in *Danske Digtanalyser* (T. Bredsdorff, ed.), pp. 135–42.
13 On land use and the concept of nature of Blicher, Aakjær, *et al.*, see Oksbjerg, E. 1977. *Naturbegrebet*, pp. 95–124.
14 On Aakjær and contemporary attitudes to the heath and to nature preservation, see Struckmann *et al.* (1943), *De Danske Heder*, 2 vols. For a more general overview of the relationship between the arts and attitudes to the heath, see Heimbürger (1969). *Den Jydske Ørken.*
15 On the history of Danish nature preservation, see Madsen (1979), *Naturfredningssagens Historie i Danmark.*
16 Aakjær (1919b), p. 474.
17 Aakjær (1919c), p. 476.
18 Aakjær (1919b), p. 490.
19 Madsen (1979), pp. 44–53.
20 Struckmann (1943). De Store Hedefredninger: in *De Danske Heder* (Struckmann *et al.*, eds), pp. 349–402. For a cartographic representation of the extent of heath preservation at present, see Dahl (1981), *Fredede Områder i Danmark.*

21 Raunkiær (1930), *Hjemstavnsfloraen hos Hedens Sangere, Blicher og Aakjær.*
22 Madsen (1979), pp. 69–95.
23 Jensen (1976), *Opgivne og Tilplantede Landbrugsarealer i Jylland*, Atlas Over Danmark Serie II, vol. I, pp. 16–17. See also Jensen (1963). A change in land-use in central Jutland, pp. 130–45.
24 Christensen (1981), *Status over Hedeplejemetoder.*
25 Jensen (1976), p. 43.
26 Oksbjerg (1977), pp. 95–124, 268–360.
27 Fullerton and Williams (1972), *Scandinavia*, pp. 111–13.
28 Hart (1955), *The British moorlands*, p. 27: Shoard (1981), The lure of the moors: in *Valued Environments* (J. Gold & J. Burgess, eds), pp. 55–73.

7 *Conclusion: nature's ideological landscape today*

The classical conception of nature has been termed an ideology because it was a conscious expression of the content of thinking characteristic of an individual, group or culture. The pastoral and agricultural stages were used in classical writings to symbolize a normal development of society and its environment. Pastoralists and their environment were identified with collective social values stemming from the nation's birth, whereas farmers and cultivated fields were associated with the values of increased reciprocity between people in society and between society and its environment. Such values were deemed vital to continued social and environmental growth. Lucretius's work, *The nature of things*, refers then to a nature conceived of as a normal process of growth and development.

Today the word nature is applied directly to the environment, to things. It is furthermore seen as landscape scenery, an esthetic entity in itself, as depicted in a painting or described by a poet. These uses of the word nature did not occur in classical times. A term which referred to a process of change was not applied to an object. Despite this transition of meaning, social values traceable to classical origins still are associated with nature. Jeppe Aakjær, for example, wished to preserve the heath not simply as physical nature, but also as a landscape identified with the common pastoral heritage of Denmark. Simultaneously and contradictorily, he appealed for its preservation as wilderness, the pre-social landscape of the solitary individual. The Heath Society of Aakjær's day, meanwhile, continued to appeal to patriotic values that identified with agriculture in its effort to create a perfect cultivated landscape of family farms interspersed with protective forest and shelter belts. Aakjær argued that the Society was more concerned with the landscape than with the social conditions of those who lived in it. Subsequent abandonment of farmlands that had been reclaimed from the heath corroborated his argument.

To identify natural values with a thing such as landscape is to objectify those values and mask or mystify their social origins. The identification of nature with a process of growth and development is removed, and hence its historicity. Thus it becomes conceptually possible to preserve environments resulting from past forms of land use (or misuse) as natural landscape – thereby the values attached to that landscape are also preserved. Human values no longer arise from a process of development in social and environmental relations, rather they seem to be posited by physical nature alone. We continue to deal with nature's ideological landscape, but the ideology is attached to a thing rather than to a social process.

It is now possible to conceive of preserving nature in landscape. In practice this is difficult to achieve. Preservation of heath from human interference as wild landscape has resulted in the growth of an unintended scrubland as plant species spread into the heath from outside the reservation and as the ecological balance within the area changes. This bushscape neither supports heathland flora and fauna nor the feeling of being alone under the mile-vaulted horizons that Aakjær described. If, however, one attempts to stabilize the heath environment by artificial methods, then the mask of wildness is lifted. The modern Danish nature preservation movement, as a consequence, is marked by heated discussion of how natural it is to tend nature by artificial means.

The attempt to preserve cultural landscape raises the same sort of problems as the preservation of wilderness. Jeppe Aakjær implies that it is possible to preserve the inherited collective values identified with the heath as pasture by preserving "this remainder of seasoned ancient earth," this shadow "on our mother's weatherbitten cheek". The problem is that the preservation of such landscape ultimately requires the preservation of a system of peasant farming which is no longer economical. The difficulty of preserving living agricultural systems is illustrated by the examples of abandoned agricultural land in modern Jutland, despite legislative restrictions. The alternative of paying farmers to cultivate and graze the land in the traditional manner would not only be costly, it would also raise difficult issues concerning the most natural means of tending nature. Would it be more natural, for example, if the farmers were forced to dwell in heather-thatched cottages with smoke holes instead of chimneys, as in the past, or should they continue to live in modern homes?[1]

The ideology of modern reclamation

Nature preservation tends to mask both physical and social conditions not only within the preserved area, but also outside the reservation. This is because the landscape becomes divided into areas designated for preservation and areas where change may occur. The campaign to save the heaths did serve to limit the Heath Society's engagement in reclamation. It did not, however, threaten the existence of the Heath Society. The society, in fact, eventually became an active partner in preservation – the Kongenshus Memorial Park, a monument to those who reclaimed the heath, is set in a heather landscape, carefully preserved by periodic burning and clearing.[2] The preservation of heathland as nature, at a time when it is only the heaths least suitable for reclamation that remain untouched, does not threaten the Heath Society; quite the opposite, this improves its public image. Now

Figure 7.1 (a) An unregulated heathland stream; (b) a regulated stream on reclaimed heath.

that nature is safely preserved the Heath Society can continue full steam with more profitable agricultural reclamation projects in unprotected areas. While preservationist attention has been focused on the heath's wide horizons, the Heath Society has been engaging in the massive drainage of wetlands and the regulation of watercourses throughout Denmark. Never have its "cold rectangular principles" been more manifest than in the once-meandering streams that have now been straightened as if they were laid out along the edge of a ruler.

The Heath Society's drainage activities have an impact on most of

Denmark's open land. It is responsible for approximately 80 percent of all drainage projects today. Half of all agricultural land has already been drained and the Ministry of Agriculture, advised by the Heath Society, is presently planning for the drainage of about 30 000 acres (12 000 ha) per year and has budgeted 40 million kroner to subsidize between one-third and one-fourth of the annual cost.[3] The effect of these activities is difficult to measure. They increase agricultural production by expanding the acreage under intensive cultivation, but recent reports have raised questions about their environmental impact. It is suspected that excessive drainage has exacerbated drought problems, as well as stream pollution, by rapidly removing water, and with it pesticides, fertilizers and other pollutants from the soil. It is normally difficult to isolate drainage as a cause of pollution, but it is occasionally fairly certain that this is the case. A celebrated example is a recent massive fish-kill at a fish nursery, which resulted from excessive amounts of ocher in the water. Ocher is leached from the soil as a consequence of draining, and the Heath Society had just completed the drainage of the fields immediately above stream. It was subsequently ruled, following the argument of the Heath Society, that ocher pollution was not covered by the Danish environmental protection legislation because it is a "natural" pollutant.[4] Although draining tends to be out of view and its impact difficult to measure, the consequences of the Heath Society's regulation of streams and rivers is quite obvious. The largest project involved the recently completed regulation of the Skjern river over a 25-km stretch inland from the river's mouth in Ringkøbing fjord on the west coast of Jutland. The project involved the construction of five large pumping stations, dikes and a parallel drainage canal for the purpose of draining nearly 10 000 acres (4000 ha) of meadow and swamp. The estimated annual 3500 tonnes of ocher which the straightened river then began to dump into the fjord were not, however, part of the original plan, nor was the subsequent damage to the fjord's fishing industry.[5]

The Heath Society's present-day reclamation efforts go relatively unchecked despite the fact that the ecological impact and even the economic rationality of its operations have been seriously questioned.[6] The Heath Society has become a national institution, a wing of the agricultural establishment with prominent politicians and businessmen on its board. Though modern agriculture is capital-intensive big business, heavily mortgaged to banking and feedstuff interests, it is still managed by the family farmer, who still seeks to maintain his image as the curator of the countryside and rural values. The fact that reclamation purports to aid the economically threatened farmer thus legitimizes the vast public subsidies which are poured into the effort. Attempts by the Danish environmental protection agency to regulate the destruction of wildlife habitats in wetlands outside nature preserves cannot measure up against the pressures for agricultural intensification.

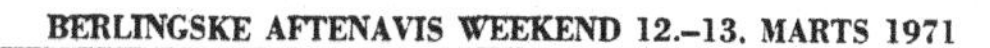

Figure 7.2 A political cartoon from the *Berlingske Weekendavis* (March 12–13, 1971) by Gerda Nystad. The caption reads: "The Heath Society is Coming!" "A stream can take care of itself for a thousand years by cleaning itself in its quiet bends. But it must be straightened, even if it violates bio-logic. This is the opinion of the Heath Society" (cartoon courtesy of Gerda Nystad).

Nature as esthetic and recreative resource

The division of the country into preserves and non-preserves tends to isolate the majority of the population from close contact with nature. The marginal agricultural lands, least disturbed by human development, are likely to be in areas distant from population centers. The Danish population has always been concentrated in areas of rich soil. Near urban areas, on the other hand, the high value of land, high wage rates and ready access to markets increases the pressure for agricultural intensification.[7] The attendant removal of hedgerow and path, drainage of meadowland and bog, and stream regulation eliminate access routes into the countryside. Agrarian landscape ceases to be an esthetic and recreative resource for

Figure 7.3 The headquarters of the Heath Society, Viborg.

human growth and development. Even forest lands under modern intensive forestry lose much of their appeal. Conifer plantations, which are the most profitable because of their high growth rate, are dark and devoid of ground vegetation and a varied fauna. Deciduous stands, when they are allowed to remain, are planted in close rows which result in stunted crowns and limited ground vegetation. Under these circumstances nature becomes something which is visited in preserves on special occasions rather than something which is readily accessible. It becomes a curiosity to be studied and photographed, or interpreted by specially trained guides. It ceases to be natural – to be in nature.[8]

International parallels

The Danish case is not unique. In Britain a recent study by Marion Shoard has documented the ways that modern environmental policy has had similar effects in segmenting the nation's environment.[9] In an article on the moors she writes:

> in Britain, the moors have captured a unique place in the imagination of many members of the countryside establishment. So much so, that the idea of protecting moorland has dominated countryside policy-making for almost the whole of the post-war period – at the expense of other types of landscape whose need has been greater. All ten of our national parks, for instance, have been selected to enshrine moorland, even though the official criteria for park designation suggest that remote moorland is far from the ideal candidate for national park status. While these parks were being designated, Britain's traditional lowland countryside – the patchwork quilt of fields, woods, downs

> and marshlands, separated by hedgerows, banks and winding streams – was undergoing a mounting onslaught from agricultural change. This lowland countryside is England's most distinctive landscape type, and survey evidence suggests it is the type most popular with the general public.[10]

Despite these preservation measures, the areas protected as nature are dwindling rapidly. The North York Moors National Park has lost nearly 60 square miles (155 sq. km) of moor between 1950 and 1975, while one-fifth of Exmoor has been plowed over since the war.[11] The loss to cultivation in parks and similar protected areas is largely due to inclusion of private agriculture within the preserved area. It has proved difficult to preserve traditional agricultural systems that maintain the moor landscape when they are no longer profitable and there is increasing pressure to cultivate the land. In areas of lowland heath preservation where agriculture has not been permitted, the British experience the same problem with bush and shrub growth which has spoiled the character of much Danish heath.

In the United States, the same bifurcation of the land into preserves and non-preserves can be observed. The Army Corps of Engineers, a rough equivalent of the Heath Society (Dalgas belonged to the Danish Army Corps of Engineers) is allowed to make massive, ecologically questionable, environmental changes in the interest of agriculture while great pains are taken to preserve wilderness enclaves in the virgin state.[12] Recent American awareness of the consequences of this division for the rural environment has created an interest in the British policy of creating parks which include agricultural land.[13] However, given the British experience, one wonders if such measures can succeed. In the United States, as in Denmark, preservation has also had unintended consequences. By protecting nature from the periodic brush fires and grazing which normally would create open meadows and remove the underbrush, the danger of major forest fires is increased at the same time as the esthetic and recreative value of a number of Western wilderness parks has been reduced. It is difficult to reverse such a policy, and recent attempts to allow fires to burn have aroused controversy.[14]

Possibly the most destructive consequences of the division of the land into preserves and non-preserves can be observed in the undeveloped countries of the Third World. In East Africa, for example, wilderness parks have been created which prohibit traditional native hunting. In consequence the population of game species, particularly the elephant, has risen dramatically. The resulting vegetation destruction by swollen herds not only has obliterated the habitat of many other wild species, it has ultimately destroyed the elephant's own resource base. Owing to the fact that elephants do not themselves distinguish between nature preserves and non-preserves, the elephant's subsequent search for food has had a consequential and destructive effect on native agriculture.[15] Having been forced off the best portions of their territory and deprived of their livelihood, the native tribes have faced material deprivation and social disintegration.[16] The major beneficiaries of these policies are the outside interests who exploit the parks for tourism.

Tourism is also the main beneficiary from the natural park development studied by Karen Fog Olwig in the West Indies. When Rockefeller financial interests purchased approximately half of the 12 000 acres (4800 ha) of the US Virgin Island of St John and donated it to the United States national park system they were careful to retain an enclave including excellent beaches. They then developed a luxury tourist paradise within what effectively became a scenic enclosure maintained by the park. Indeed more than a scenic enclosure, it protected the resort from outside competition because the park controlled the island's best beaches. The protection of so much of the island from human productive use effectively removed the traditional native resource base, which depended on large areas of territory for grazing and extensive cultivation. This in turn stimulated a movement of the native population into the island's port village, where they formed an inexpensive labor reserve for the developing tourist industry. Today the native economy is severely hampered by the fact that less than 3000 acres (1200 ha) are left accessible to use. Furthermore, competition for this land with outside buyers has resulted in an enormously inflated price of between 30 000 and 40 000 dollars per acre (12 000–16 000 dollars per ha). Meanwhile the land under preservation is growing into an impenetrable bushscape, making a mockery of park officials' dreams of restoring the former sugar plantation and peasant lands to a supposed "pre-Columbian" forest.[17]

Finis

By regarding nature as a thing, a landscape, we deem it capable of preservation; in practice such preservation requires that the entire landscape be frozen, because society and its environment are continuously interacting and changing.[18] Physical and ecological processes do not respect the boundaries of nature reserves. Cultivated species and manufactured chemicals spread into nature areas while wild species can rarely exist entirely within the bounds of a preserve. Landscapes created by past forms of environmental use, or lack of use, will probably cease to exist in their present form if they are preserved from human contact. The most perfectly formed modern agricultural landscape will deteriorate if the material basis of its existence within a changing society is ignored. The same applies to the values which are identified with these landscapes. One does not preserve collective values that provide the basis for national existence by preserving a pastoral landscape and, since the landscape is ever-changing, one cannot even preserve in it the memories of those values. Likewise, we are not able to preserve the ideal of reciprocity identified with traditional agriculture, by creating a perfect landscape of field and wood, if that agriculture is based on the extractive exploitation of the environment's marginal yield by a farmer, who is himself desperately struggling to pay high rates of interest on borrowed capital. We cannot preserve values of individualism in a wild landscape when the maintenance of its wildness depends on a complex industrial society, whose existence is predicated upon the division of labor. Vacation paradises are no garden of Eden if their preserved flora and fauna cannot survive within their boundaries, or if the natives of the area are deprived of human dignity and independence. Under such circumstances nature ceases to be natural.

Notes

1 On the problem of preserving historical landscapes, see: Lowenthal and Binney (1981), *Our past before us, why do we save it?*; Newcomb (1979), *Planning the past: historical landscape resources and recreation.*
2 Skodshøj (1953), *Hedens Opdyrkning i Danmark.*
3 Geckler (1982), *Hvad Indad Tabes, Hedeselskabets Virksomhed, Magt og Position*, pp. 115–17.
4 Ibid., pp. 71–94.
5 Ibid, pp. 9–45.
6 Ibid, pp. 145–61.
7 Two recent studies of the problem in an international context are: Bryant *et al.* (1982), *The city's countryside: land and its management in the rural–urban fringe*; and Furuseth and Pierce (1982), *Agricultural land in an urban society.*
8 I have treated this problem in more detail in Olwig, K. (1983), Natur, Langtidsplanlægning og Utopi på Sjælland, pp. 28–36. See also Frykman and Löfgren (1979), *Den kultiverade människan*, pp. 1–27.
9 Shoard (1980), *The theft of the countryside.*
10 Shoard (1982), The lure of the moors, p. 55.
11 Ibid., p. 71.
12 On the impact of this dichotomization, see: Brandt (1977), Views, pp. 46–9; Lowenthal (1964), Is wilderness "paradise enow"?, pp. 34–40; Runte (1979), *National parks: the American experience*; and Shepard (1967). *Man in the landscape.*
13 Shoard (1980), p. 152.
14 Cooper (1961), The ecology of fire, pp. 150–60: Vankat (1977), Fire and man in Sequoia National Park, pp. 17–27.
15 Beard (1977), *The end of the game*: Graham (1973), *The gardeners of Eden.*
16 Turnbull (1972), *The mountain people.*
17 Olwig, K. F. (1980), National parks, tourism, and local development: a West Indian case, pp. 22–31.
18 On problems and ambiguities involved in preserving the past in general, see: Lowenthal (1976), Past time, present place: landscape and memory, pp. 1–36; and Lowenthal (1979), Age and artifact, dilemmas of appreciation, pp. 103–28.

Bibliography

Aagesen, Aa. 1943. Fra Hedegaard til Bysamfund. In Struckmann *et al.* (1943, pp. 263–98).

Aakjær, J. 1903–1904. *Steen Steensen Blichers Livstragedie i Breve og Akt-Stykker*, 3 vols. Copenhagen: Gyldendal.

Aakjær, J. 1915. *Hedevandringer*. Copenhagen: Gyldendal.

Aakjær, J. 1918. Paa Hedens Høje. In *Samlede Værker*, vol I, 211–14.

Aakjær, J. 1919a. Heden. In *Samlede Værker*. vol. V: Af *Min Hjemstavns Saga*, 599–710.

Aakjær, J. 1919b. Hedens Fredning. In *Værker*, vol. IV, 479–90.

Aakjær, J. 1919c. Naturfredning (orig. 1909). In *Værker*, vol. IV, 474–78, Copenhagen: Gyldendal.

Aakjær, J. 1919d. Vredens Børn. In *Værker*, vol. VI, 439–650.

Abbey, E. 1971. *Desert solitaire: a season in the wilderness*. New York: Ballantine.

Andersen, H. C. 1951. *Mit Livs Eventyr*, 2 vols, H. Topsoe-Jensen (ed.). Copenhagen: Gyldendal.

Andersen, H. C. 1964. Jylland Mellem Tvende Have. In *Folkehøjskolens Sangbog*, 15th edn. Odense: Forening for Højskoler og Landbrugsskoler, no. 530.

Andersen, V. 1970. *Den Jyske Hedekolonisation*. Aarhus: University Press.

Auden, W. H. 1967. *The enchafed flood, or the romantic iconography of the sea*. New York: Random House, Vintage.

Baggesen, S. 1965. *Den Blicherske Novelle*. Copenhagen: Gyldendal.

Barfod, F. 1853. *Fortællinger af Fædrelandets Historie*. Copenhagen: Gyldendal.

Barrell, J. 1972. *The idea of landscape and the sense of place 1730—1840: an approach to the poetry of John Clare*. Cambridge: Cambridge University Press.

Barrell, J. 1980. *The dark side of the landscape: the rural poor in English painting 1730—1840*. Cambridge: Cambridge University Press.

Barrell, J. and J. Bull (eds) 1974. *The Penguin book of English pastoral verse*. London: Allen Lane.

Barthes, R. 1972. *Mythologies*, A. Lavers (transl.). New York: Hill & Wang.

Beard, P. H. 1977. *The end of the game*. Garden City, New York: Doubleday.

Bendix, R. 1901. *C. Dalgas, Hans Liv og Kunst*. Copenhagen: Jacobsen.

Bisgaard, H. L. 1902. *Den Danske National Økonomi i det 18. Aarhundrede*. Copenhagen: H. Hagerup.

Bjerregaard, H. 1840. *Er den Evige Fred en Saadan Chimære og Umulighed, som en Æret Forfatter Nylig har Yttret*? Randers: J. M. Elmhoff.

Blair, H. 1803. A critical dissertation on the poems of Ossian the son of Fingal. In Macpherson, 1803, 49–155.

Blicher, N. 1978. *Topographie af Vium Præstekald*. Herning: Poul Kristensen.

Blicher, S. S. 1823. *Bautastene*. Odense: Hempel.

Blicher, S. S. 1920a. *Steen Steensen Blichers Samlede Skrifter*, J. Aakjær and H. Ussing (eds), vols. I and II: *Ossians Digte*. Copenhagen: Gyldendal.

Blicher, S. S. 1920b. Bedømmelse over Skrivtet Moses og Jesus. In *Skrifter*, J. Aakjær and H. Ussing (eds), vol. III, 34–90.

Blicher, S. S. 1920c. Hiemvee. In *Skrifter*, J. Aakjær and H. Ussing (eds), vol. III, 130–7.

Blicher, S. S. 1920d. Jyllandsrejse i Sex Døgn. In *Skrifter*, J. Aakjær and H. Ussing (eds), vol. IV, 120–229.

Blicher, S. S. 1924a. Danmarks Nuværende Tilstand. In *Skrifter*, J. Aakjær and G. Christensen (eds), vol. XIII, 170–89.
Blicher, S. S. 1924b. Digterens Lyksalighed. In *Skrifter*, J. Aakjær and G. Christensen (eds), vol. XIV, 122–4.
Blicher, S. S. 1927. Eneboeren på Bolbjerg. In *Skrifter*, J. Aakjær and G. Christensen (eds), vol XIX, 123–50.
Blicher, S. S. 1928. Viborg Amt. In *Skrifter*, J. Aakjær and J. Nørvig (eds), vol. XXI, 199–268, vol. XXII, 1–240.
Blicher, S. S. 1930a. De Tre Helligaftener. In *Skrifter*, J. Aakjær and J. Nørvig (eds), vol. XXVI, 1–15.
Blicher, S. S. 1930b. Min Tidsalder. In *Skrifter*, J. Aakjær and J. Nørvig (eds), vol. XXVI, 153–265.
Blicher S. S. 1931. Bordtale ved Himmelbjergfesten 1843. In *Skrifter*, J. Nørvig (ed.), vol. XXVII, 2–11.
Blicher, S. S. 1933. Letter to Peter Christian Koch. In *Skrifter*, J. Nørvig (ed.), vol. XXXII, 163–4, Copenhagen: Gyldendal.
Blicher, S. S. 1945a. The hosier and his daughter. In *Twelve stories*, H.A. Larsen (transl.), 220–39. Princeton, NJ: Princeton University Press.
Blicher, S. S. 1945b. The robber's den. In *Twelve stories*, 79–121.
Brach, C. H. 1879. *Om Molesworths Skrift, 'An account of Denmark as it was in the year 1692'*. Copenhagen: C. H. Reitzel.
Brandt, A. 1977. Views. *Atlantic Monthly* (July 1977).
Bredsdorff, T. 1975. *Digterens Natur*. Copenhagen: Gyldendal.
Brostrøm, T. 1969. Digt og Digterjeg, Jeppe Aakjær: Sundt Blod. In *Danske Digtanalyser*, T. Bredsdorff (ed.), Copenhagen: Arena. 135–42.
Brown, J. C. 1878. *Pine plantations on the sand wastes of France*. Edinburgh: Oliver & Boyd.
Bryant, C. R., L. H. Russwurm and A. G. McLellan 1982. *The city's countryside: land and its management in the rural urban fringe*. London: Longman.
Burnet, T. 1965. *The sacred theory of the Earth* (orig. 1684). London: Centaur Press.

Carstens, F. C. 1839. *Bemærkninger over Alheden og dens Colonier*. Viborg: P.C. Kabell.
Carstens, F. C. 1844. *Bemærkninger over Heden og dens Træplantning*. Copenhagen: J.S. Schubother.
Christensen, P. G. 1981. *Status over Hedeplejemetoder*. Copenhagen: Fredningsstyrelsen.
Collingwood, R. G. 1976. *The idea of nature*. Oxford: Oxford University Press.
Cooper, C. F. 1961. The ecology of fire. *Scient. Am.* **CCIV**, 150–60.
Cour, J. C. la 1865. Landbruget i Kempen (Campinen). *Tidsskrift for Landoekonomie* Ser. 3, **XIII**, 217–85.
Cour, J. C. la 1866. Om Hedernes Benyttelse. *Tidsskrift for Landoekonomie* Ser. 3, **XIV**, 141–6.
Curtius, E. R. 1953. *European literature and the Latin middle ages*, W. R. Trask (transl.). London: Kegan Paul.

Dahl, K. 1981. *Fredede Områder i Danmark*. Copenhagen: Danmarks Naturfredningsforening.
Dalgas, E. 1865. Om Opdyrkning af de Jydske Heder. *Berlingske Politiske og Avertissements-Tidende*, no. 287 (Wed., Nov. 15), no. 288 (Thurs. Nov. 16).
Dalgas, E. 1866. *En Oversigt over Hederne i Jylland*. Aarhus: Wissing.
Dalgas, E. 1867–8. *Geographiske Billeder fra Heden*. Copenhagen: C. A. Reitzel.
Dalgas, E. 1873. Det Jydske Haveselskab. *Ugeskrift for Landmænd* **V**, 4 række, 366–9.

Dalgas, E. 1875. *Anvisning til Anlæg af Smaaplantninger omkring Gaarde og Haver samt til Anlæg af levende Hegn og Anlæg af Pileculturer.* Copenhagen: Hedeselskabet.
Dalgas, E. 1884. Fortids- og Fremtids- Skovene i Jyllands Hedeegne. *Hedeselskabets Tidsskrift* **IV**, no. 5 (July 1883), 135–40; no. 6 (Aug., Sept. 1883), 153–69; no. 9 (Dec. 1883), 211–24. **V**, no. 1 (Jan., Feb., Mar. 1884), 1–72 (illus. map). **VI**, no. 3 (June, Jul., Aug. 1885), 87–110, no. 4 (Sept., Oct. 1885), 123–40; no. 5 (Nov., Dec. 1885), 146–60.
Dalgas, E. 1887. *Hedesagens Fremgang*: 1866–87. Aarhus: Hedeselskabet.
Dalgas, E. 1891. *Familien Dalgas: Slægtsregister fra 1685 til 1891 med Beretninger om Familiens Medlemmer og mine egne Livserindringer.* Aarhus.
Dalgas, E. 1903. *Sangbog*, Axel Mielche (ed.). Copenhagen: Lehman og stage.
Daniels, S. 1982. Humphrey Repton and the morality of landscape. In *Valued environments*, J. R. Gold and J. Burgess (eds.), 124–44. London: George Allen & Unwin.
Dessau, D. 1866. *Den tiende danske landmandsforsamling i Aarhus 25–29 Juni 1866.* Copenhagen: Triers Bogtrykkeri.

Eliot, T. S. 1944. *What is a classic?* London: Faber & Faber.
Erichsen, L. M. 1903. *Den Jydske Hede Før og Nu.* Copenhagen: Gyldendal.
Everhart, W. C. 1972. *The national park service.* New York: Praeger.

Farley, F. E. 1903. *Scandinavian influences in the English romantic movement*, Studies and Notes in Philology and Literature IX. Boston: Ginn.
Feingold, R. 1978. *Nature and society.* Hassocks, Sussex: The Harvester Press.
Forchhammer, G. 1835. *Danmarks geognostiske Forhold, forsaavidt som de ere afhængige af Dannelser, der ere sluttede, fremstillede i et Indbydelsesskrift til Reformationsfesten den 14de Novbr. 1835.* Copenhagen: Jens Hostrup Schulz.
Forchhammer, G. 1844. Om en stor Vandflod, der har truffet Danmark i en meget gammel Tid. *Dansk Folkekalender for 1844*, F. C. Olsen (ed.), 84–96. Copenhagen: Selskabet for Trykkefrihedens rette Brug.
Forchhammer, G. 1855. De Jydske Heder. *Dansk Maanedeskrift* **I**, 161–80.
Frye, N. 1971. *Anatomy of criticism.* Princeton, NJ: Princeton University Press.
Frykman, J. and O. Löfgren 1979. *Den kultiverade människan.* Lund: Liber Läromedel.
Fullerton, B. and A. F. Williams 1972. *Scandinavia.* London: Chatto & Windus.
Furuseth, O. J. and J. T. Pierce 1982. *Agricultural land in an urban society.* Washington, DC: Assoc. Am. Geogs.

Garboe, A. 1961. *Geologiens Historie i Danmark*, vol. II. Copenhagen: C. A. Reitzel.
Geckler, R. 1982. *Hvnd Indad Tabes: Hedeselskabets Virksomhed, Magt og Position.* Copenhagen: Gyldendal.
Glacken, C. J. (1956) Changing ideas of the habitable world. In *Man's role in changing the face of the Earth*, W. L. Thomas (ed.), 70–92. Chicago: University of Chicago Press.
Glacken, C. J. 1967. *Traces on the Rhodian shore: nature and culture in Western thought from ancient times to the end of the eighteenth century.* Berkeley: University of California Press.
Goldschmidt, M. 1954. *En Hedereise i Viborg-Egnen*, V. Andersen (ed.). Kolding: Konrad Jørgensen.
Goldsmith, O. 1773. *The deserted village.* Leipzig: Altenburgh.
Graber, L. H. 1976. *Wilderness as sacred space.* Monog. Ser. no. 8. Washington, DC: Assoc. Am. Geogs.
Graham, A. D. 1973. *The gardeners of Eden.* London: George Allen & Unwin.

Graham, I. C. C. 1956. *Colonists from Scotland: emigration to North America, 1707—1783*. Ithaca, New York: Cornell University Press.
Grøn, A. H. 1933. *Hvad Nytte er Skoven Til?* Copenhagen: Levin & Munksgaard.

Hansen, V. 1970. Hedens Opståen og Omfang. In *Danmarks Natur*, A. Nørrevang and T. J. Meyer (eds). Vol. VII: *Hede, Overdrev og Eng*. Copenhagen: Politiken.
Hansen, V. 1971. Rural settlement on Danish glacial outwash plains. *Les congres et colloques de L'Universite de Liegen*. Vol. LVIII: *L'Habitat et les paysages ruraux d'Europe*. Liege: Universite de Liege.
Hard, G. 1965. Arkadien in Deutschland; Bemerkungen zu einem landschaftlichem Reiz. *Die Erde* **XCVI**, 21–41.
Hart, J. F. 1955. *The British moorlands*. University of Georgia Monogs, no. 2. Athens, Georgia: University of Georgia Press.
Heimbürger, M. 1969. *Den Jydske Ørken*. Copenhagen: Grevas.
Henning, F. W. 1972. Die grosse Stadt in verschiedenen Verhältnissen betrachtet (J. H. G. v. Justi 1764). *Zeitschrift für Agrargeschichte* **XX**, 186–97.
Hoffman, H. de 1757. *Oeconomiske betragtninger om Aarhus-Stift*. Copenhagen: Nicolaus Møller.
Hoffman, H. de 1758. Om Heederne i Jylland. Heedens Dyrknings Forbedring. *Oeconomisk J.* (Jan.), 20–46.
Hoffman, H. de 1768. *Den Danske Atlas*, vol. IV. Copenhagen: A. H. Godiche.
Hoffman, H. de 1781. *Samtale Angaaende Hedernes Dyrkning og Forbedring i Jylland, som et Anhgang til Hr. Conferentz-Raad og Amtmand Fleischers Agerdyrknings Cathechismus Hvilken det Kongelige Landhuusholdnings Selskab har ladet trykke og uddeele*. Odense.

Ingemann, B. S. 1815. Til Ossians Fordansker, Steen Steensen Blicher, I Anledning af Hans Første Udgivne Digte. *Dansk Minerva* **I** (Oct.), 338–41.

Jensen, K. M. 1963. A Change in Land-Use in Central Jutland. *Geografisk Tidsskrift* **LXIII**, 130–45.
Jensen, K. M. 1976. *Opgivne og Tilplantede Landbrugsarealer i Jylland*, Atlas Over Danmark Ser. II, vol. I. Copenhagen: Reitzel.
von Justi, J. H. G. 1758 [1760]. Allerunterthänigstes Gutachten, Wegen Anbauung der jütländischen Heiden. Keine Oberfläche der Erden ist Schlechterdings und von sich selbst unfruchtbar. In *Neue Wahrheiten zum Vortheil der Naturkunde und des gesellschaftlichen Lebens der Menschen* **XII**, 609–79. Leipzig: Bernhard Christoph Breitkoph.

Kierkegaard, S. 1906. Af en Endnu Levendes Papirer. In *Samlede Værker*, A. B. Drachmann, J. L. Heiberg and H. O. Lange (eds), vol. XIII, 41–92. Copenhagen: Gyldendal.
Kliger, S. 1945. The 'Goths' in England: an introduction to the Gothic vogue in 18th-century aesthetic discussion. *Modern Philology* **XLIII** (2), 10–16.
Kliger, S. 1947. The Gothic revival and the German translation. *Modern Philology* **XLV** (2), 73–103.
Kliger, S. 1952. *The Goths in England*. Cambridge, Mass.: Harvard University Press.

Leiss, W. 1974. *The domination of nature*. Boston: Beacon Press.
Lewis, C. S. 1967. *Studies in words*, 2nd edn. Cambridge: Cambridge University Press.
Lovejoy, A. O. 1923. The supposed primitivism of Rousseau's discourse on inequality. *Modern Philology* **XXI**, 165–86.

Lovejoy, A. O. 1927. 'Nature' as aesthetic norm. *Mod. Lang. Notes* **XLII** (7), 444–50.

Lovejoy, A. O. 1973. *The great chain of being*. Cambridge, Mass.: Harvard University Press.

Lovejoy, A. O. and G. Boas 1935. *Primitivism and related ideas in antiquity*. Baltimore: Johns Hopkins Press.

Lowenthal, D. 1964. Is wilderness 'paradise enow'? Images of nature in America. *Columbia Univ. Forum* **VII** (2), 34–40.

Lowenthal, D. 1976. Past time, present place: landscape and memory. *Geogr. Rev.* **LXV**, 1–36.

Lowenthal, D. 1979. Age artifact, dilemmas of appreciation. In *The interpretation of ordinary landscapes* (D. Meinig, ed.). Oxford: Oxford University Press.

Lowenthal, D. and M. Binney (eds) 1981. *Our past before us, why do we save it?* London: Temple Smith.

Lucretius 1951. *On the nature of the universe*, R. E. Latham (transl.). Harmondsworth: Penguin.

Lukács, G. 1962. *The historical novel*, H. Mitchell and S. Mitchell (transl.). London: Merlin Press.

Lund, N. J. W. 1975. Kartoffeltyskerne. *Folk og Kultur: Årbog for Dansk Etnologi og Folkemindevidenskab*, 31–66. Copenhagen: Foreningen Danmarks Folkeminder.

Macpherson, J. 1773. *Introduction to the history of Great Britain and Ireland*, 3rd edn, London: T. Becket and P. A. De Hondt.

Macpherson, J. 1803. *The poems of Ossian, the son of Fingal* (orig. 1765). Edinburgh: Denham & Dick.

Madsen, F. K. 1979. *Naturfredningssagens Historie i Danmark*. Odense: Odense Universitetsforlag.

Mallet, P. H. 1756. *Indledning udi Danmarks Riges Historie* (transl. anon.). Copenhagen: Ludolph Heinrich Lille.

Mallet, P. H. 1770. *Northern antiquities: or a description of the manners, customs, religion and laws of the ancient Danes and other northern nations including those of our own Saxon ancestors*, 2 vols, (T. Percy anon. transl.). London: Carnan.

Marsh, G. P. 1843. *The Goths in New England*. Middlebury, Vermont: Philomathesian Society of Middlebury College.

Marx, K. 1973. *Grundrisse*, M. Nicolaus (transl.). Harmondsworth: Penguin.

Marx, L. 1964. *The machine in the garden*. New York: Oxford University Press.

Matthiessen, H. 1939. *Den Sorte Jyde*. Copenhagen: Gyldendal.

McLaren, M. 1970. *Sir Walter Scott, the man and patriot*. London: Heinemann.

Molbech, C. 1829. Optegnelser paa en Udflugt i Jylland i Sommeren 1828. *Nordisk Tidsskrift for Historie, Litterarur og Kunst* **III**, 106–76.

Molesworth, R. 1694. *An account of Denmark as it was in the year 1692*, 3rd edn. London: Timothy Goodwin.

Møller, P. L. 1971 (original 1847). *Steen Steensen Blicher, Kritiske Skizzer fra Aarene 1840*—7, 173–82. Copenhagen: Gyldendal.

Muir, E. 1936. *Scott and Scotland*. London: Routledge.

Nash, R. 1973. *Wilderness and the American mind*, revised edn. New Haven, Conn.: Yale University Press.

Newcomb, R. 1979. *Planning the past: historical landscape resources and recreation*. Folkestone, Kent: Dawson.

Nicolson, M. H. 1963. *Mountain gloom and mountain glory*. New York: W. W. Norton.

Nørgaard, F. 1941. *Jeppe Aakjær. En Introduktion til Hans Forfatterskab*. Copenhagen: Gyldendal.
Nørvig, J. 1943. *Steen Steensen Blicher, Hans Liv og Værker*. Copenhagen: Gyldendal.

Oakley, S. 1972. *A short history of Denmark*. New York: Praeger.
Ogden, J. T. 1974. From spatial to aesthetic distance in the eighteenth century. *J. Hist. Ideas* **XXXV** (1), 63–78.
Oksbjerg, E. 1977. *Naturbegrebet*. Copenhagen: Rhodos.
Olufsen, C., Captain Selmer and Landinspecteur Kynde 1800. Efterretninger om Alheden og Randbølheden i Nørre Jylland. *Økonomiske Annaler* **III** (1), 84–98.
Olufsen, C. 1811. *Danmarks Brændselsvæsen. Physikalskt, Cameralistisk og Oekonomisk Betragtet*. Copenhagen.
Olwig, K. 1974. Place, society and the individual in the authorship of St. St. Blicher. In *Omkring Blicher 1974*, F. Nørgaard (ed.). Copenhagen: Gyldendal.
Olwig, K. 1980. Historical geography and the society/nature 'problematic': the perspective of J. F. Schouw, G. P. Marsh and E. Reclus. *J. Hist. Geog.* **VI** (1), 29–45.
Olwig, K. 1981a. Var Blicher den Vilde Hedes Eller Menneskets Digter? In *Museerne i Viborg Amt*, Marianne Bro-Jørgensen (ed.) vol. XI, Viborg: Viborg Stiftsmuseum.
Olwig, K. 1981b. Literature and 'reality': the transformation of the Jutland heath. In *Humanistic geography and literature*, D.C.D. Pockock (ed.), 47–65. London: Croom Helm.
Olwig, K. 1983. Natur, Langtidsplanlægning og Utopi på Sjælland. *Geografisk Magasin* **CXXX**, 28–36.
Olwig, K. F. 1980. National parks, tourism, and local development: a West Indian case. *Human Organization* **XXXIX** (1), 22–31.
Oppermann, A. 1889. *Bidrag til det Danske Skovbrugs Historie 1786—1889*. Copenhagen: Gyldendal.

Panofsky, E. 1955. *Meaning in the visual arts*. New York: Doubleday Anchor.
Parry, A. 1957. Landscape in Greek poetry. *Yale Classical Studies* **XV**, 3–29.
Passmore, J. 1974. *Man's responsibility for nature*. New York: Charles Scribner's Sons.
Pedersen, E. H. 1971. *Hedesagen under forvandling: Det Danske Hedeselskabs historie 1914—1966*. Copenhagen: Gyldendal.
Pontoppidan, E. (ed.) 1757. Introduction. *Oeconomiske Magazin* **I**, I–XVI.
Pontoppidan, E. (ed.) 1761. Introduction. *Oeconomiske Magazin* **IV**, I–XII.
Pontoppidan, E. 1763. *Den Danske atlas*, vol. I. Copenhagen: A. H. Godiche.
Putnam, M. C. J. 1970. *Virgil's pastoral art*. Princeton, NJ: Princeton University Press.

Rasmussen, P. and N. W. Lund 1974. *Hedekolonierne*. Godsarkiver 2, Arkivregistraturer 4. Viborg: Landsarkivet for Nørrejylland.
Raunkiær, C. 1930. *Hjemstavnsfloraen hos Hedens Sangere, Blicher og Aakjær*. Copenhagen: Schultz.
Rhode, P. P. 1967. Idédigtning og Politisk Gennembrud. In *Dansk Litteratur Historie*, P. H. Trausted (ed.). Vol. II: *Fra Oehlenschläger til Kierkegaard*, 595–701. Copenhagen: Politiken.

Riegels, H. C. 1847. Brudstykker fra en Udflugt over Hannover, Brunsvig og Magdeborg, især med Hensyn til Træplantning og Havecultur. *Tidsskrift for Landoekonomie*, new series, **VIII**, 76–128.

Riegels, H. C. 1848. *Til Træplantnings Fremme i Almindelighed dog fornemmelig paa Hede- og andre slette Jorder der kun egner sig til Skovcultur*. Copenhagen: S. L. Møller.

Riis, J. A. 1910. *Hero tales of the far north*. New York: Macmillan.

Runte, A. 1979. *National parks: the American experience*. Lincoln, Neb.: University of Nebraska Press.

Ruskin, J. 1904. Of classical landscape. In *The works of John Ruskin*, E. T. Cook and A. Wedderburn (eds). London: George Allen.

Schelde, N. 1761. Frie-Tanker om Aarsagerne til Folkemangel i Danmark. *Oeconomiske Magazin* **V**, 23–56.

Schmidt, A. F. 1933. *Jeppe Aakjær, Nogle Oplysninger om Hans Forfatterskab*. Copenhagen: Gyldendal.

Schouw, J. F. 1852a. Nature and nations. In *The earth, plants and man*, 240–6.

Schouw, J. F. 1852b. The action of the human race upon nature. In *The earth, plants and man*, 228–39. London: Bohn.

Scott, W. 1863. *The Highland clans* (orig. 1816). Edinburgh: Adam & Charles Black.

Shepard, P. 1967. *Man in the landscape*. New York: Ballantine.

Shoard, M. 1980. *The theft of the countryside*. London: Temple Smith.

Shoard, M. 1982. The lure of the moors. In *Valued environments*, J. Gold and J. Burgess (eds), 55–73. London: George Allen & Unwin.

Skodshøj, H. (ed.) 1953. *Hedens Opdyrkning i Danmark*. Viborg: Det Danske Hedeselskab.

Skodshøj, H. 1966. *E. M. Dalgas*. Viborg: Det Danske Hedeselskab.

Skrubbeltrang, F. 1966. *Det Indvundne Danmark*. Copenhagen: Gyldendal.

Small, A. W. n.d. (orig. 1909). *The Cameralists: the pioneers of German social polity*, Burt Franklin Research and Source Works Series, no. 43. New York: Burt Franklin.

Snell, B. 1953. *The discovery of the mind*. Cambridge, Mass.: Harvard University Press.

Struckmann, E. 1943. De Store Hedefredninger. In Struckmann *et al*. 1943, 349–402.

Struckmann, E., K. Jessen and F. Hjerl-Hansen (eds) 1943. *De Danske Heder: Deres Natur og Fortidsminder, Folkeliv og Kultur*, 2 vols. Copenhagen: Arthur Jensen.

Tegner, E. 1817. Svea. *Svenska Akademiens Handlingar ifrån År 1796*, no. 6., 155–70.

Testrup, S. 1759. Forslag om Hederne i Nørre-Jylland til Ager og Eng at optage. *Oeconomiske Magazin* **III**, 91–112.

Thomson, D. S. 1951. *The Gaelic sources of Macpherson's 'Ossian'*. Edinburgh: Oliver & Boyd.

Thorpe, H. 1957. A special case of heath reclamation in the Alheden district of Jutland, 1700–1955. *Trans. Paps Inst. Br. Geogs* no. 23, 87–121.

Tombo, Jr., R. 1966. *Ossian in Germany, bibliography, general survey: Ossian's influence upon Klopstock and the bards*. New York: AMS Press.

Tuan, Y. F. 1971. *Man and nature*. Resource Paper no. 10. Washington, DC: Assoc. Am. Geogs.

Tuan, Y. F. 1974. *Topophilia: a study of environmental perception, attitudes and values.* Englewood Cliffs, NJ: Prentice-Hall.

Tuan, Y. F. 1976. Geopiety: a theme in man's attachment to nature and to place. In *Geographies of the mind,* D. Lowenthal and M. J. Bowden (eds), 11–39. New York: Oxford University Press.

Turnbull, C. M. 1972. *The mountain people.* New Haven, Conn.: Yale University Press.

Turner, J. 1979. *The politics of landscape.* Cambridge, Mass.: Harvard University Press.

Turnock, D. 1970. *Patterns of Highland development.* London: Macmillan.

Vankat, J. L. 1977. Fire and man in Sequoia National Park. *Annls Assoc. Am. Geogs* **LXVII**, 17–27.

Varro 1960. *On agriculture,* W. D. Hooper (transl.), H.B. Ash (revised). Cambridge, Mass.: Harvard University Press.

Virgil 1946. *Eclogues and Georgics,* T. F. Royds (transl.). Everyman's Library 222, revised edn. London: Dent.

Webster's seventh new collegiate dictionary 1963. Springfield, Mass.: G&C Merriam.

Whitney, L. 1924. English primitivistic theories of epic origins. *Modern Philology* **XXI** (4), 337–78.

Willey, B. 1962. *The eighteenth-century background.* Harmondsworth: Penguin Peregrine.

Williams, R. 1972. Ideas of nature. In *Ecology the shaping enquiry,* J. Benthall (ed.), 146–64. London: Longman.

Williams, R. 1973. *The country and the city.* New York: Oxford University Press.

Williams, R. 1976. *Keywords.* London: Fontana, Croom Helm.

Index

Printed in the United States
by Baker & Taylor Publisher Services